C000156295

© 2016 Camden Miniature Steam Services

British Library Cataloguing-in-Publication-Data:
a catalogue record of this book is held by the British Library.

First Printing 1914 by Percival Marshall & Co. London

ISBN No. 978-1-909358-35-5

Published in Great Britain by:
CAMDEN MINIATURE STEAM SERVICES
Barrow Farm, Rode, Frome, Somerset. BA11 6PS
www.camdenmin.co.uk

Camden stock one of the widest selections of fine transportation,
engineering and other books; contact them at the above address, or see their website,
for a copy of their latest free Booklist.

PUBLISHER'S NOTES:

Our grateful thanks to Dennis Sharp of Hailsham for his help in the reprinting this book.

The original does not have a publication date, but on pages 14 & 15 reference is made to a boat club obtaining a club house at the beginning of 1914. This, together with the styles of the boats depicted, and the complete absence of any reference to hostilities, would suggest a publication date just before the outbreak of World War One.

Safety: whatever its date of publication, the reader should remember that safety standards one hundred years ago were much more lax than today, and should interpret instructions in this book accordingly. Safety of yourself and others, both in the workshop, and when running your boat, are entirely your responsibility, and the publishers cannot accept any responsibility or liability for this.

PLEASE NOTE:

Due to the slightly marked original copy from which most of this book was scanned, there are marks on certain pages which have been 'carried-over' from the original, which we regret we have been unable to entirely eliminate.

A Model Power Boat Regatta

MODEL
POWER BOATS

A Complete Manual on the Designing, Building,
and Running of all kinds of Model
Self-propelled Boats, Steam,
Petrol and Electric.

BY

E.W. Hobbs

Associate of the Institution of Naval Architects.
Member of the Junior Institution of Engineers

FULLY ILLUSTRATED

CONTENTS.

LIST OF PRINCIPAL DESIGNS.

PREFACE.

THE object which I have had in view in preparing this book has been twofold—firstly, to demonstrate the magnitude of the subject, its history and development; and secondly, to show that Model Boating has necessarily advanced beyond the stage of a mere game, and has attained the status of an organized sport, involving scientific and technical considerations. A consideration of the scientific side of the subject also leads in an easy and natural manner to a deeper knowledge and understanding of Naval Architecture—a subject at all times of paramount importance to Great Britain. All the dwellers within these shores are dependent upon the sea, her ships, and seamen, for the very necessaries of life themselves; and some knowledge of the fundamentals of Naval Architecture—when presented in an attractive manner—should be of more than passing interest.

The early history of Model Boats and their subsequent development occupy the first chapter, while the second is devoted to a comprehensive survey of ships in general, and the special features to be embodied in a successful model. Next comes a consideration of the theoretical aspect of the subject, and explanations are given of how and why a boat floats, and the various forces acting on a ship are dealt with. A further chapter shows how to utilise these technicalities in the formation of an harmonious and complete design.

The " practical man " may be inclined to scoff at these technics, but he should always remember that theory is not merely an epitome of the hardly-gained experiences of the past, but serves the purpose of a lighthouse in pointing out to the unwary navigator the whereabouts of the rocks and shoals, for which reason those who seek success in the handling of their boats should make theory their servant and perseverance their colleague.

The construction of Hulls, of various types, is detailed fully, and some instruction given in fitting out a model ; while separate chapters deal in as comprehensive a manner as space permits with such items as Boilers, Burners, Petrol and Electric Motors, and other mechanical details. Space necessarily restricts the scope of this part of the work, but the author hopes sufficient has been given to indicate successful lines for work.

As the character of a model depends so much upon the deck fittings and superstructures, it has been thought well to devote a whole chapter to them, and some hundred odd fittings are enumerated, the pages being interspersed with illustrations of typically well-finished models. For the would-be racing expert an explanation of the rules under which Model Power Boats are raced is given, together with some hints on how to manage the boat, while the concluding chapter comprises a glossary of technical and nautical terms which will possibly be of assistance to some readers. The general idea has been to present the subject from—as far as possible—a popular standpoint, and it is hoped that the book will be of use to readers in suggesting new models for construction, or improvements to existing vessels.

In conclusion, the author takes this opportunity of expressing his thanks to numerous kind friends who have tendered information and photos, and given encouragement in the preparation of this volume. It is impossible to name them all, but the author's best thanks are given with all heartiness and sincerity, and it is hoped that this work may be of as much help to others as the work of others has been to the author. After all, progress can only be made by appreciating the good work of others, and doing one's best to improve thereon and offer to the newcomers a helping hand.

My special thanks must, however, be accorded to Messrs. Bassett-Lowke, Ltd., of London and Northampton, who have given freely of information, photographs, and illustrations, and have permitted the use of their technical information.

To my friend Mr. Bassett-Lowke—who is well known as a photographer of great ability, apart from his engineering attainments—a special meed of thanks is due, as his work in taking many of the photographs specially for me has materially assisted in the illustration of the book.

The assistance of my friends Mr. A. H. Avery, Mr. Wintering-ham, Mr. Mackenzie, and Messrs. Stuart Turner, Ltd., must not be forgotten. While for much information reference has been made to the classical works of Prof. Sir J. Biles, Mr. Seaton, Mr. Dixon Kemp, and the Transactions of the I.N.A. And finally I have to thank Mr. Percival Marshall for his unremitting care in editing these pages.

EDWARD W. HOBBS.

LONDON, S.W.

INTRODUCTION.

AMONGST the earliest recorded achievements of mankind none rank higher than those of the shipwright, as the desire of man to voyage on the face of the waters has always been a trait of those nations living on the sea-coast or adjacent to the great rivers. There have been found in the old Egyptian tombs models of ships entombed with the embalmed bodies of young Egyptian princes, deceased three thousand years B.C., and these models were possibly used as playthings, and were probably very common in those days.

Subsequently, as humanity advanced in knowledge of the liberal arts and sciences, as well as in the claims of empire, so ships increased in accuracy of design and value. The Romans constructed some of the finest examples of the shipwright's art, and their handicraft has been exemplified by a few relics of ships of that period that have been found from time to time. In more recent days, the great naval architect, if so we may term him, " Peter the Great," used models to demonstrate his ideas of ship building. Although in the Middle Ages the construction of models was not practised to any extent, there are in existence one or two old models of some interest ; while in comparatively modern times quite a wealth of ship models may be found illustrating clearly an advanced state in the science of naval architecture. About the beginning of the seventeenth century the development of sailing ships,consequent upon the expansion of trade, was very marked, and the museums both in England and upon the Continent have many fine examples of models of these vessels. The application of models of yachts and ships for serious educational and sporting purposes was only fully realised within the last decade, although there have existed since about 1840 one or two model yacht clubs, in London and the provinces, but it has remained for the twentieth century to produce clubs in numbers devoting their energies to the sporting side of model yachting.

From an educational standpoint the classical experiment with ship models, carried out about 1740 by Col. Beaufoy, marked a new era in naval architecture, whilst the colossal results achieved by the later experiments of the elder Froude with ship models must always rank as the greatest contribution from any one man to the science of naval architecture ; and to the scientist who proved beyond doubt the "law of comparison" between a ship model and its parent form, the thanks of all naval architects are due ; and in conclusion he was the means of producing the first experimental or testing tank for model boats in England, this tank being built at the naval station of Haslar about 1850.

An interesting old warship model is reproduced in Fig. 1. This is a scale model of the *Majestic*, an early British battleship. The model—except the rigging—is entirely constructed of cardboard, and was built by Mr. B. Audsley, an American model maker, the boat now being on exhibition in the rooms of the Yacht Club at Mauritius.

It will be seen that science and the highest traditions of our nation are allied with the use and construction of model boats, and it is hoped that in these pages readers will find something of interest and instruction on this all-important subject of ships and shipping : a subject which must always be of vital interest to any nation with a seaboard, and above all an island kingdom such as Great Britain.

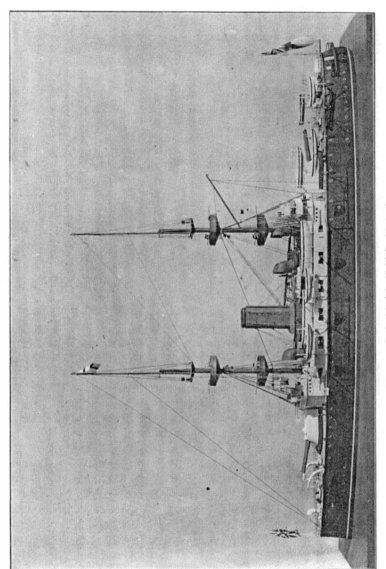

Fig. 1. Scale Model, H.M.S. *Majestic*.

Model Yacht Club Section at the *Model Engineer* Exhibition, 1913.

CHAPTER I.

IT is only in recent years that the possibilities of model boating as a sport have been partially appreciated, although the interest which is now being taken in all forms of model boats cannot but be regarded as a healthful sign of the nation's growing recognition of the all-important matter of maritime power. Added to patriotic reasons, there are the wonderful educative possibilities of a good scale model, while as a social or sporting factor model boating claims serious consideration. No one who has been present at a club meeting can have failed to notice the general spirit of *camaraderie* which exists among the members. The intense interest of the bystander watching the earlier processes of steam-raising and preparatory work, the breathless suspense and exhilarating excitement of the race, often with humorous results, render the pastime both fascinating and absorbing. Regarded as a serious sport, there is undoubtedly a great stimulus to one's inventive faculties in the desire to possess a crack racer and one which, for at least a time, holds the supreme position in her class. Such a result cannot, of course, be obtained without considerable knowledge of the subject, and some general experience with the handling of the model, and to secure this end there are few finer advantages than those obtained by joining one of the numerous model boat clubs. Readers of the *Model Engineer* will no doubt be familiar with the recently increased interest, on the part of the model yacht clubs and the model engineering societies, in model steamers and power boats. Notices are fairly frequent of the formation of new clubs, while space is often occupied with photos and descriptive notes of model regattas and speed boat races. Probably the *Model Engineer* Speed Boat Races against time, held annually, receive the largest patronage, and the winner of this race may well consider his boat a " crack racer," as some of the best in England take part.   The races are run at any convenient time and place, but have to be properly checked and timed by fully qualified model yachtsmen or members of a known society of model engineers.   The results of the 1913 *Model Engineer* Speed Boat Competition are :—

Class A.
> *Jan III* (J. Andriaensen).   32½ lbs. displacement. Speed, 10.75 m.p.h.   Silver Medal.

Class B.
> Petrol Boat H.P. (T. P. Mears).   25 lbs. displacement. Speed, 13.11 m.p.h.   Silver Medal.

Class C.
> *Bulrush* (G. D. & S. S. Noble).   15 lbs. 2 ozs. displacement.   Speed, 22.77 m.p.h.   Silver Medal.

Class D.
> *Irene II* (H. H. Groves).   6 lbs. 13½ ozs. displacement. Speed, 21.19 m.p.h.   Silver Medal.
>
> *Savage* (G. D. & S. S. Noble).   9 lbs. 15 ozs. displacement. Speed, 17.18 m.p.h.   Bronze Medal.
>
> *Tiddley* 2 (G. D. & S. S. Noble).   5 lbs. 10 ozs. displacement.   Speed, 16.38 m.p.h.   Certificate.
>
> *Rattler* (A. Rankine).   9.5 lbs. displacement.   Speed, 6.88 m.p.h.   Certificate.

The figures given above need but little comment ; whilst to the record breakers themselves we feel sure no congratulations are needful to promote a pleasantly peaceful consciousness of well-earned victory.

One of the most go-ahead power boat clubs now in London is the Victoria Model Steamer Club.   Races are conducted by this club practically all the year round, and visitors may always expect a warm welcome.   The Club possesses its own boat-house, and the use of a fine lake at Victoria Park, London, E.   There are also strong clubs in other parts of London, such as Forest Gate Power Boat Club, the Wimbledon Model Power Boat Club, the Kensington Model Steamer Club, and the North Middlesex Model Steam Navigation Club.

The Forest Gate Club possesses about twenty-five boats, and a number of others are being built.   At the beginning of the year

1914 the Club was fortunate enough to obtain a club house, and as it is quite near the lake it will prove a boon to the members for storing their boats and holding those very useful impromptu meetings to discuss model steamer matters.

The North of England is well represented in Liverpool, and Scotland by the Glasgow Model Steamer Club, and others. In the South and West of England there are also several prominent clubs, notably the Portsmouth Model Steamer Club, which was founded in 1910. Its headquarters are at the Canoe Lake, Southsea. The President is Commander J. F. Grant-Dalton, R.N., and the Vice-Presidents are Percival Marshall, Esq., of the *Model Engineer*, Fred. T. Jane, Esq., the famous naval expert, and Stuart Turner, Esq., of engineering fame. The Club was one of the first to become

Fig. 2.   *Maude II* and the P.M.Y.C.
Championship Shield.

affiliated to the M.Y.R.A., and has met with a fair amount of success in the National Regattas.

The illustration Fig. 2 shows the $1\frac{1}{2}$ metre Class B racer, *Maude II*, owned by Mr. H. E. Burden, of Gosport. Speed, 10 m.p.h., two Stuart s/a engines arranged tandem, driving one propeller. The Scott boiler used is fired by petrol blow lamp. Hull, bright cedar and varnished. The shield was presented to the P.M.S.C. by the President, Commander J. F. Grant-Dalton, R.N., and is a splendid specimen of the silversmith's and enameller's art.

Other model power boat clubs are springing into being very rapidly. The work of the Model Yacht Racing Association, the recognised head and ruling body of the sport, is now well known. This Association holds regattas annually for speed boats and sailing models, at which solid silver challenge cups, medals, and many other handsome prizes are awarded. The competition at these regattas is very keen, model boating enthusiasts coming from all parts to witness and take part in the racing.

Not only does the Model Yacht Racing Association arrange regattas, but it promotes indoor meetings, with the view of improving matters of design, etc., assists at exhibitions, and generally promotes the improvement of model yachting. The M.Y.R.A. Stand at the

Fig. 3. A Display of Ship Models by the M.Y.R.A. at Olympia.

recent Children's Welfare Exhibition, promoted by the *Daily News and Leader*, and held at Olympia, London, is shown in Fig. 3, and gives a good idea of the popularity of model boats and power boating. On p. 12 is shown a group of power boats at the *Model Engineer* Exhibition, October, 1913, where these models formed a most important section of the exhibits.

One of the best movements of the M.Y.R.A. has been the arrangement of International Regattas for both sailing and power boats. In 1913 this was held at Enghien les Bains, a very charming suburb of Paris. The first day was devoted to sailing yachts, the power boats being raced on the following day. A full description of this interesting event was published in the *Model Engineer* of October 2nd, 1913.

Fig. 4 shows the English team, while Fig. 5 illustrates the International Cup and some of the prizes.

Fig. 4.   International Model Yachting.   The English Team at Paris, 1913.

We will now proceed with some consideration of the models generally in use amongst the devotees of this sport.

As regards the type of boat usually adopted, this is naturally largely a matter of individual taste, governed by the object to be obtained. For purely racing purposes, a petrol or steam vessel of extreme type, shorn of all unnecessary fittings, becomes essential; but for the mere enjoyment of a pleasant pastime, a vessel more in the nature of a scale model usually finds favour, and it is with particular reference to such models that the following paragraphs have been written. The chief consideration in the selection of a model power boat not intended for purely racing purposes is its scale. By reference to Fig. 6 some idea of the relative proportions of three typical vessels,

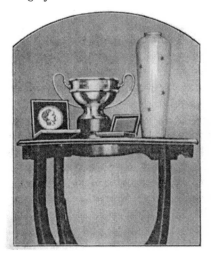

Fig. 5.   The International Cup and the President's Vase.

constructed in the model to a uniform length, can be obtained.   The
tug boat, A, is built to a scale of approximately $\frac{1}{2}$ in. to the foot, and
represents a vessel 85 ft. long.   The torpedo boat destroyer, B, would
reasonably be supposed to measure 170 ft. long in the prototype,
and if the model were built to the same size as the tug boat (3 ft. 3 in.,
or one metre long), she would approximate to a scale of $\frac{1}{4}$ in. to
the foot, or exactly one-half the scale of the tug boat, whereas, if we
take an extreme type, and build a model of the *Mauretania*, C,
only 3 ft. 3 in. long, it would have a scale of approximately $\frac{1}{20}$ in.
to the foot, so would be one-fifth of the scale of the torpedo boat destroyer, from which it must be fairly obvious that the fittings of the
tug boat are larger and can be more easily and accurately modelled.
The torpedo boat destroyer makes a model of reasonable size, but
possesses several disadvantages, which will be gone into later,
while it would be practically impossible to build a good scale model
of the *Mauretania* of so small a length, and if it were otherwise, no useful purpose would be served by building a model
so small.   If a model is intended for a good scale appearance,
and capable of long and steady working, vessels of the tug boat type
should be selected, that is, the model should be built on the lines
of a smaller-sized prototype, leaving the larger prototypes for large
scale models, in preference to modelling a large vessel to a small
scale, as this invariably leads to intricate and minute working, or
the character of the vessel will inevitably be lost.   Taking all these
features into consideration, and the added difficulty of making a
very small engine and boiler, there is little doubt that the basis for
the smallest practical model is the *Metre*, a model measuring
39 in. over all.   With such a length quite a number of reasonable
and practical models can be built.   For cruising purposes the one
and a half metre, or 4 ft. 6 in., size may be selected as a useful length.
Thus a similar tug boat to Fig. 6, A, would measure approximately
4 ft. 6 in. long, being about $\frac{3}{4}$ in. to the foot scale, and the T.B.D.
$\frac{3}{8}$ in. to the foot scale, but care must be taken not to make the model
too heavy in the latter size, or difficulty will be experienced in
carrying it about.   Models 5 ft. 6 in. and over in length are
frequently adopted for large vessels such as liners and battleships,
but with this size the difficulty of transportation becomes very
great, and the use of a boat-house adjacent to the water is therefore
extremely desirable.   Of course, there are many conditions under

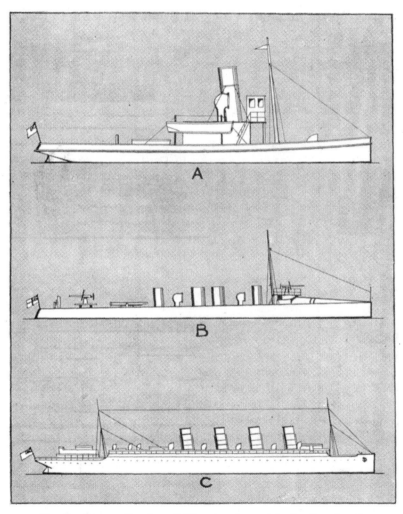

Fig. 6.   Comparative sizes of different-scales models of uniform length.

which such models should be selected. They are ideal sizes for practical use on open waters, the only disadvantage being the trouble, already mentioned, of taking them from place to place. When this condition does not apply, the 5 ft. 6 in. model, or even a larger size, should receive serious consideration. Such models are frequently built, but usually only where they can be housed near the water upon which they are to be sailed, or otherwise suit the individual requirements of the builder.

For racing purposes, two sizes of boat are practically universal, the M.Y.R.A. Class A and Class B. These are restricted to certain overall dimensions. There are also the M.Y.R.A. Unrestricted Classes C, D, and E, while the *Model Engineer* holds annual contests for boats limited only to displacement as follows :—

Class A	..	..	.. 10 lbs.
Class B	..	..	.. 20 lbs.
Class C	..	..	.. 30 lbs.
Class D	..	..	.. 40 lbs.

The majority of the club matches are now raced under M.Y.R.A. Rules and for its Classes A and B; the particulars of these M.Y.R.A. Rules are as follows : —

Class A.

The length overall shall not exceed one metre (39¾ in.), beam shall not exceed 20 per cent of the length, and the displacement in proper working order with all fuel and water on board shall not exceed 12 lbs.

Class B.

The length overall shall not exceed one and a half metres (59 in.), beam shall not exceed 20 per cent of the length, and displacement shall not exceed 25 lbs. in proper working order with all fuel and water on board.

These two classes are without doubt the most popular; they produce a fast, reliable boat, that of the Class A being exceptionally good, speeds of 14 to 16 m.p.h. being frequently obtained. The Rules under which these events are held and hints on the running of the boats will be found in Chapter XIV.

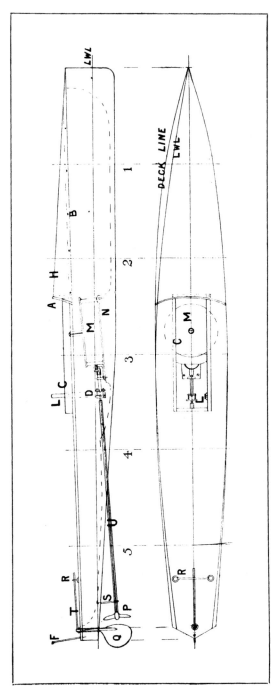

Fig. 7. Design for small Clockwork Motor Boat.

CHAPTER II.

THE examples of power boats to be described in this chapter are intended to show the general characteristics and features of the most practical prototypes. The question of motive power is dealt with later in succeeding chapters, where details are given of the machinery mentioned in reference to the particular type of craft being described. The smallest type of model can conveniently be driven by clockwork, and such vessels are very suitable for children or for experimental work, although they are not practical in sizes over one metre in length. Clockwork provides the cleanest and most reliable of all motive powers available. Fig. 7 shows the design of a small clockwork racing motor boat, 24 in. long, 3 in. beam, and 2 in. in depth. This little model can readily be constructed by an amateur, and if carefully made will attain a speed of about $1\frac{1}{2}$ m.p.h. The clockwork motor M is obtainable from advertisers in this book, and is mounted on a block of wood N. A " driver " is fitted at D, between the motor and the propeller shaft U, which is supported by a wire skeg or bracket S. The propeller P is $1\frac{1}{4}$ in. dia., three bladed. A plain rudder Q, with tiller T, and rack R, controls the boat's direction. The mast F carries the house flag, while the lever L stops and starts the motor. A plain deck with coaming C, around the hatch or opening, with a spray hood H forwards, edged with the half round brass wire at A, and screwed to the hull side at B, completes the little model.

Fig. 8 shows a smart river launch 30 in. long, $5\frac{1}{4}$ in. beam, and $3\frac{1}{4}$ in. depth, capable of attaining a speed of 2 m.p.h., with clockwork mechanism, although a very small electric motor and accumulator might be used in its place, when the speed would be greater.

This design for a model up-river launch is somewhat out of the ordinary run of model power boats, but the type is so handsome and

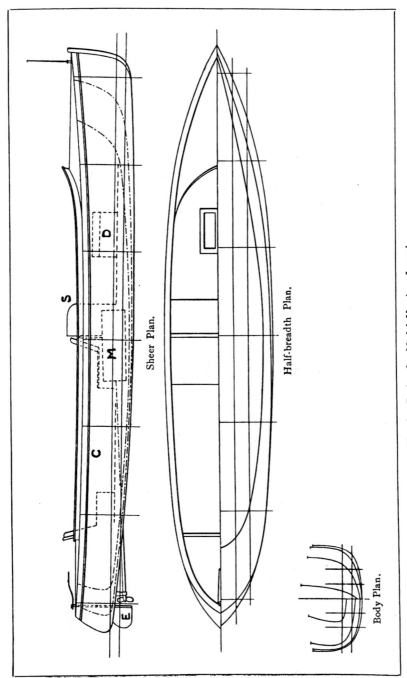

Sheer Plan.

Half-breadth Plan.

Body Plan.

Fig. 8. Design for Model Up-river Launch.

distinctive that it is well worth the serious consideration of model
power boat builders who are not possessed purely with the idea of
" speed at all costs." The design lends itself more particularly
to the construction of the small boat, and the writer had in mind a
model measuring 30 in. in length, $5\frac{1}{4}$ in. beam, and 1 in. draft, this
size being very suitable for propulsion by powerful clockwork
motors. The motor M is arranged amidships, immediately under
the engineer's seat S, and the forward seats in the cockpit C.

The prototype is arranged for petrol propulsion, the motor being
housed in the casing D, in the forward part of the boat, but in
the model the clockwork motor is brought further aft for the sake of
stability and trim, as well as from the fact that in this position it is
entirely hidden from view. The hull could either be cut from the
solid or built up in the bread and butter fashion. If this latter
method is adopted, and it is usually considered the best, it will
require one plank 2 in. thick and one plank 1 in. thick. A slight addition to the thickness of this plank will have to be made at the bows
of the boat to allow for the sheer. Two transverse bulk-heads should
be fitted at the commencement fore and aft of the coaming, while
to obtain the best effect in the model the deck should be planked
with narrow planks $\frac{1}{4}$ in. in width, alternate mahogany and pine,
with a side strake of mahogany. The hull should be finished with
white enamel, and below the water line should be finished with a
dark or anti-fouling green. The narrow bead or rubbing strake
could very well be finished gilded. As regards the internal fixings
of the model the amount of detail that is to be put into the boat
depends entirely upon the individual ideas of the builder. The
seats should be neatly covered with red plush, and a neat grating
on the floor should certainly be fitted, as well as a flag mast and
mooring bollard forwards.

Fig. 9 shows a fine open cruising launch, such as seen for harbour service. The accumulator is carried forward under the hood,
and the motor in the cockpit, where it is encased in an engine cover.
The smallest practical size for a steam driven vessel is about 24 in.
long, and Fig. 10 gives a useful design for such a vessel. This represents an ordinary class of cabin cruiser which lends itself admirably
to steam propulsion. The hull may be cut from clean dry pine,
and decked with $\frac{1}{8}$ in. pine, finished in white and green. The propelling machinery consists of a simple brass boiler and plain spirit

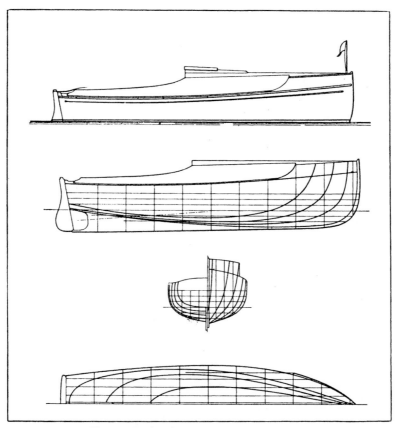

Fig. 9.   Design for a Model Cruising Launch.

lamp. The engine is of the oscillating type, and drives a small three-bladed propeller; such little models are very interesting and much fun can be obtained with them.

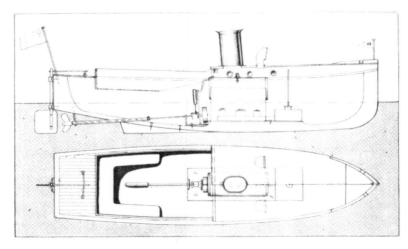

Fig. 10.   Simple Model Steam Launch.

An improved 24 in. Class E racing motor boat is shown in Fig. 11. This is fitted with twin screws; the boiler of copper, with three $\frac{1}{4}$ in. water tubes, is arranged forward, the lamp, of the open pan spirit type, having asbestos wick, and a spirit reservoir for wards;

Fig. 11.   24 in. Steam-driven Racing Boat.

a single cylinder engine $\frac{5}{16}$ in. bore $\frac{1}{2}$ in. stroke drives the two propellers through a double slide crank and gearing, making a fast light boat. The difficulties inherent in the construction of a small simple steamer are very considerable, although the design suggested in Fig. 12 is of a practical model. This is a metre size torpedo boat destroyer, of a more elaborate type with an enlarged boiler, vaporizing spirit lamp, and improved engine. This boat would have a speed of 4 to 5 m.p.h., when everything was running satisfactorily. For an all-round useful steam-driven model the steam yacht is hard to beat, and a good example of such a model is given in Chapter XII. Fig. 13 is an illustration of a simple cargo boat, which is one of the easiest types of model to construct. Deck fittings are reduced to the minimum, while the prototype itself is fairly clear of fittings, and consequently the

Fig. 12. A Steam-driven Model T.B. Destroyer.

model can be readily made to give a good scale effect. These vessels are almost invariably painted black above the water line, and red below, and are usually driven with a single screw, and this arrangement has been adopted in the model of the *Wentworth*, Fig. 13, which is a typical example of what may be done in the way of producing a simple, effective, and cheap model.

On the raised forecastle at the bows of the boat are stowed two stockless anchors in the hawse pipes, the anchor chains being led to a dummy windlass. The only other fittings on this particular boat are a small ventilator and the stanchions and railings round the side of the hull.

In the waist are situated two cargo hatches, and the foremast with derricks and booms.   These are all quite simple in construction, and details of construction can easily be seen from the illustration. Amidships are situated the officers' and passengers' accommodation. Forwards of the funnel is the chart house with navigation bridge.   Two large ventilators provide air for the stokehold ; an engine room skylight and two small ventilators provide light and air to the engine room. Of course, in the model these are only dummy, but, nevertheless, give a handsome appearance to the boat.   Four lifeboats on davits with blocks and falls are carried for service or life-saving on the boat, these being stowed away on light boat skids.   The main hatches and mainmast are similar to the foremast,

Fig. 13.   A Model Cargo Boat.

and on the poop, right aft, is shown a deck saloon, the roof of this being removable to give access to the tiller and rack which control the rudder.   The boat is finished black above the water line, and red below.   Stanchions and rails are silver plated, funnel red and black, ventilators buff, ship's boats white with brown gunwale, the chart house, etc., are white, the windows and doors carefully painted on, the result being an inexpensive, characteristic, and practical model boat, with all the more important fittings of appropriate scale size. The actual model in the photograph measured 36 in. in length, with a beam of 5 in. and total depth of 4 in.

When the construction of naval models is attempted, much difficulty will probably be experienced in retaining the correct and characteristic aspect so necessary in a scale model.   The size of the

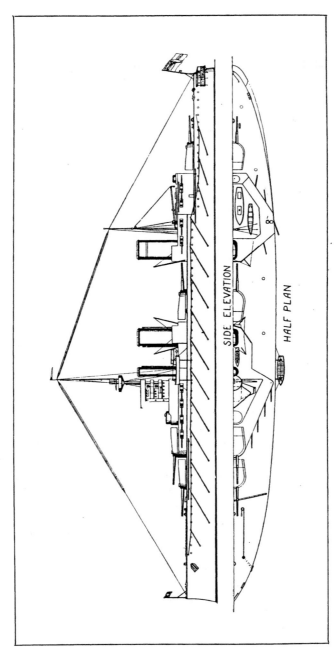

SIDE ELEVATION

HALF PLAN

Fig. 15. Deck and Sheer Plan of Model of H.M.S. *Princess Royal.*

originals renders a large model practically imperative, and there is
little doubt, in the light of experience, that such models should be
built to a scale of $\frac{1}{10}$ in. to the foot.   This makes a model
cruiser of modern type about 6 ft. 9 in. long, while a battleship on the
lines of the *Iron Duke* will be over 5 ft. long.   The scale is rather
small to adopt for a practical model of a torpedo boat destroyer,
but $\frac{1}{8}$ in. scale might be adopted and would not look materially out
of proportion.   The illustrations given show three typical warships,
Fig. 14, representing the City Class, a fast light cruiser or corvette.
Fig. 15 is characteristic of such vessels as the *Princess Royal*, a
type of boat which is being retained in the Navy and is a fast,
powerful battle cruiser.

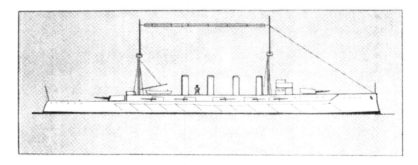

Fig. 14.   Typical Light Cruiser.

Such a model is always attractive on any pond, and the illustration
of one of the latest battle cruisers, H.M.S. *Princess Royal*, is an excellent prototype for a model on which it is desired to show a number of
deck fittings, while at the same time to retain all that is best in a
working model.   Such a boat should undoubtedly have twin screws ;
although quadruple screws would be correct in an exhibition or
glass case model, twin screws will be found more satisfactory and
reliable for ordinary work.

The salient features of the modern British battle cruiser from
bow to stern are—the heavy stockless anchors stowed in hawse
pipes ; 13.5 guns, one pair on the main deck, the other pair mounted
on a turret arranged to fire over the top of the forward pair of guns ;
immediately aft of these guns is the armoured conning tower, and

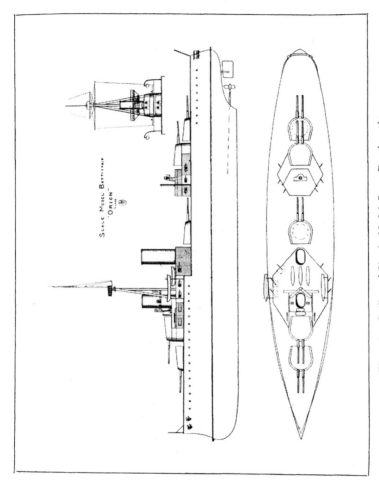

SCALE MODEL BATTLESHIP
"ORION"
CLASS

Fig. 16.   Outline of Simple Model Super-Dreadnought.

behind this the steel chart house with two or more bridges for navigation and observation purposes.   On these bridges are mounted
the searchlights, and at the back of the bridge is situated the pole
mast with its fire control station and wireless yards.   The chart
house and navigation platforms are mounted on a steel structure
known as the forward battery, in which are mounted eight 6 in. guns.
Behind the battery is the first funnel, usually oval in section.
Between the forward funnel and the main funnel which is frequently of circular section, are stowed a number of ship's boats
protected from the blast of the midship guns by steel screens.

Practically in the centre of the ship are situated a pair of 13.5
guns, and behind these is the after funnel, which is generally long
and narrow to reduce windage.   Immediately behind this funnel
is a short pole mast with a boat derrick, and arranged around this
mast, in a sort of well, are the steam pinnaces and other boats.
These are situate on top of the after battery, which contains eight
6 in. guns, an armoured conning tower being built at the extreme
after end of the battery, and behind this is still another pair of 13.5
guns.

At the extreme stern of the boat is the admiral's sternwalk or
quarter gallery.   On the outside of the hull are the torpedo booms
with net, and a well made model equipped with the fittings already
enumerated would look most handsome and realistic.   The various
fittings are described in detail in the chapter on " Deck Fittings,"
and, of course, may be applied to any vessel according to their scale
size and prototype.

Fig. 16 represents a model of the *Orion* type, one of the latest
types of modern battleships, and a development of the famous
Dreadnought.   The *King George* is shown in Fig. 17, and is one of
several models designed by the author, and built by Bassett-Lowke,
Ltd., for exhibition at Madame Tussaud's in Baker Street.   On all
these models a noticeable absence of small fittings will be observed,
but this is thoroughly characteristic of the originals, which are, as
far as possible, shorn of all unnecessary details, and therefore
form most valuable prototypes to the model maker.   Either steam
engines or electric motors may be selected for these models ; both
give excellent results.   A pleasing variation from the usual modelled
torpedo boat destroyers is found by building one of the now popular
" Scouts."   These vessels measure in the original approximately

Fig. 17. Working Model Super-Dreadnought. H.M.S. *King George V.*

360 ft. in length, and can therefore be reasonably made to a scale
of $\frac{1}{10}$ in. to the foot giving a model 36 in. in length.   These boats are
light, simple in construction, and extremely handsome in appearance,
while usually possessing quite a good speed; such a boat is shown in
outline in Fig. 18.

. Another type of vessel that is particularly appropriate for modelling
is the tug boat, and a good example is given in Fig. 19 of a tug, the
Lady Mary.   This is described in detail in the January, 1913, issue
of the handy little journal *Junior Mechanics*.

Fig. 20 is a reproduction of a photograph of a magnificent steam
yacht, the *Ituna*, and shows her in her native element on the

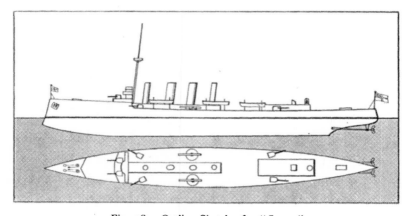

Fig. 18.   Outline Sketch of a " Scout."

Round Pond, Kensington, where she presents a noble sight.   This
fine model is the property of J. J. Daniels, Esq., Vice-President
of the Kensington Model Power Boat Club.   She measures about
9 ft. in length, is fitted up in most perfect fashion, equipped with
triple expansion engines, coal fired Yarrow boiler, and donkey pump
for the water feed.   Her speed is considerable, while the pleasing
white and gold finish gains for her owner much well merited praise.

Liners, in one form or another, have always been popular as models.
Their variety is legion, but the general characteristics are length and
a multitude of superimposed decks.   Fig. 21 is an outline in plan
and elevation of a modern two funnel boat, suitable for models of
about one metre in length; as it will be noted, the beam of the model

Fig. 20. The Model Steam Yacht, *Iluna*, on the Round Pond, Kensington Gardens.

is increased over scale width, to ensure sufficient stability and displacement.

Fig. 22 shows in outline a very handsome single funnel boat of the " White Star " type, and would make a good steaming model about 4 ft. 6 in. long. Fig. 23 is a photo reproduction of a similar model, but of a different company. This is the *Deutschland*, of the Hamburg Amerika Co. The model measures 6 ft. 6 in. long, and is replete with all details ; such a well finished model is, however, only really suited to exhibition purposes under a glass case.

Fig. 119.   The Model Steam Tug *Lady Mary*.

The submarine does not lend itself to satisfactory modelling, neither does the more modern " submersible," although an above-water profile of a modern submarine is given in Fig. 24, and the general lines of a submersible in Fig. 25. The difficulty with both these types of boat is the motive power, and clockwork appears to be the only all-round satisfactory arrangement, as it is very reliable and can be entirely enclosed, thus rendering the hull watertight.

Another type of vessel, specially suited to steam machinery, and

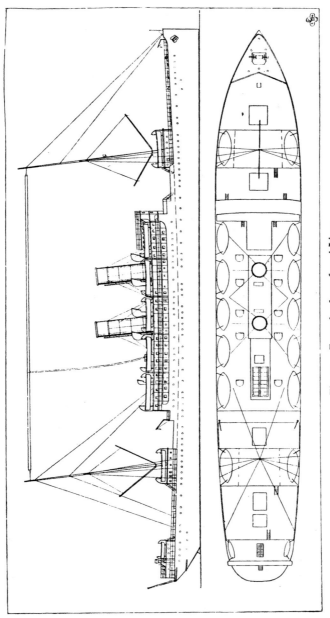

Fig. 21. Example of two-funnel Liner.

making an excellent model is shown in Fig. 26, which is a scale
model vedette boat, as used in the Navy for harbour service, and
for a hundred and one jobs by the " handy men " ; it makes an
interesting boat.   The illustration shows such a model designed
by the author for Count Gabriel de Lonyay, and constructed by

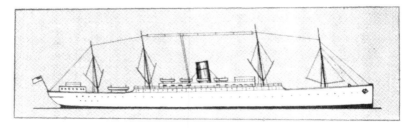

Fig. 22.   Model " White Star " Liner.

Bassett-Lowke, Ltd.   The steering wheel is arranged to operate
the rudder by means of chains and wire ropes in the orthodox manner.
The machinery consists of a Simplex engine, water tube boiler, and
petrol blow lamp.   The whole is most attractive and works exceedingly well.

Fig. 23.   Model Hamburg-American Liner.

   Electricity as a motive power for the smaller sizes of scale models
should be seriously considered and generally adopted, as it is clean
and always ready for use, provided the accumulators are kept
charged.   Such models as river launches and elaborate liners, or
any model with a large variety of detail fittings, are hardly suitable

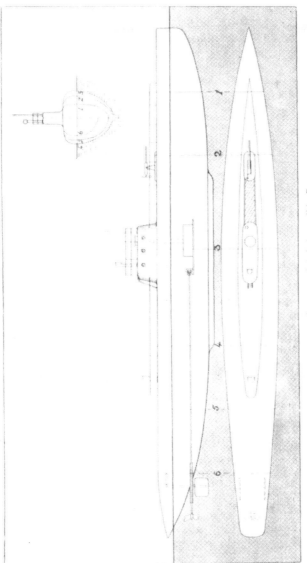

Fig. 25. Lines for a Model Submersible Boat.

for steam machinery, as the heat and water are very liable to cause damage to the paintwork and fittings, and it is here that electricity scores owing to its cleanliness. Some examples of electric boats have already been given here, but further particulars are given in Chapter II, which is devoted entirely to such models.

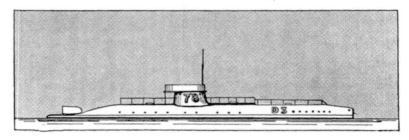

Fig. 24.   Above-water Profile of a Modern Submarine.

The photograph on page 56 shows a handsome type of torpedo boat destroyer, with a large number of deck fittings and details. The lines of a similar hull are given and the method of design is described fully in Chapter V, on "How to Design a Model Power Boat."

Fig. 26.   Model Vedette Boat.

Electricity is the most suitable motive power for such a model, as the displacement is not sufficient for steam machinery to be successfully installed.

The motor may be a No. 10 *Nautilus*, with gear box, driving two $1\frac{1}{4}$ in. diameter propellers, current being supplied from a small

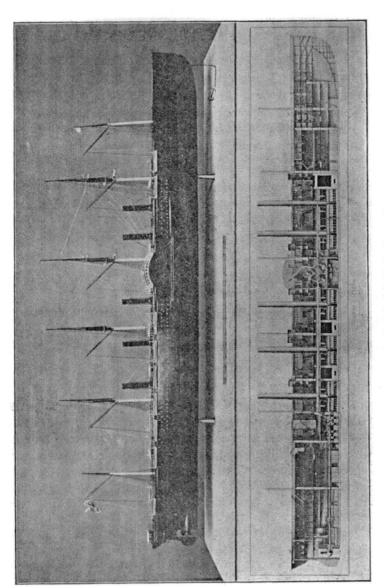

Fig. 27. Scale Model of R.M.S. *Great Eastern*.

4 volt accumulator ; but reference should be made to the chapter
on " Electric Machinery " for particulars of the method of installation and other details.

From electrically driven models to the pure exhibition model is
but a step, and although hundreds of photographs could be reproduced of such models, only three examples are given here, as it is
believed that the space can better be utilised by description of
actual practical working models ; but the fine model, now in South
Kensington Museum, of that wonderful old vessel, the *Great Eastern*,
Fig. 27, built at Millbank about 1850–1854, from the combined
efforts of Scott Russell, I. K. Brunel, and James Watt & Co., all
master minds, and engineers of the highest order, is one that has
much of interest to the student of shipbuilding. Not only was

Fig. 28.   Scale Model Cabin Cruiser.

this ship one of the largest built for many years, but it combined
many features that are recognised to-day as of paramount importance.   For instance, she had a double skin, vertical bulk-heads not
pierced with doors, longitudinal framing, and composite propulsive
machinery.   The paddle wheels are, of course, useless for modern
service conditions, but the wealth of detail shown in the design
should gain a great tribute to the foresight of her builders.   The
particulars of the ship were as follows, for she has now been broken
up :—

Length, 692 ft. ; beam, 82 ft. ; draft, 30 ft. ; displacement, 27.384 tons ; horse power, 8.297 ; speed, 14 knots.
Passenger accommodation : 1st Class, 800 ; 2nd class,
2,000 ; 3rd Class, 1,200.

Fig. 29. Clyde Steamer, *Queen Alexandra.*

While as a troop ship she could carry 10,000 men. The model is to a scale of ⅛ in. to the foot, and measures approximately 7 ft. 3 in. in length.

An altogether different type of exhibition model is reproduced in Fig. 28, and is a model of a 60 ft. cruising launch, built for J. A. Holder, Esq. The model is to a scale of ½ in. to the foot, and was built by Bassett-Lowke, Ltd. It is a most pleasing design, and would make an excellent working model with electrical machinery.

Fig. 29 represents still another type of model. This is one of the very handsome screw boats extensively used on the Clyde. She is the *Queen Alexandra* built in 1902 by Messrs. Denny & Co., of Dumbarton. The length of the actual vessel is 270 ft., and beam of 32 ft., but a fine model could be made to a scale of ⅛ in.

Fig. 31.   Typical Petrol-driven Racing Motor Boat.

to the foot. The beam should not be less than 6 in. and depth 4 in. The absence of deck fittings combined with the neat arrangement of decks, saloon, funnels, and boats being particularly pleasing ; this makes the boat most suitable as a prototype for a neat practical working model steamer.

Fig. 30 gives lines of an ordinary " service launch " hull, a good sound design, capable of moderately fast speeds.

For purely racing boats, a petrol motor of either the two or four cycle type is frequently adopted, and a typical example is shown in Fig. 31, but reference should be made to Chapter X, where the subject of " Petrol Motor Boats " is dealt with in detail. High pressure steam racing motor boats hold a premier place in many enthusiasts' opinion, and the excellent examples furnished by the recent *Model Engineer* races give some idea of the results to be obtained from a steam driven vessel. At the time of going to press several models

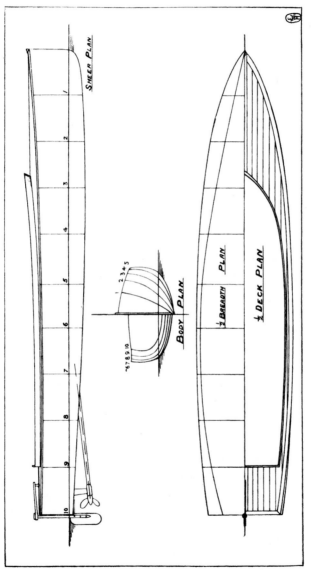

SHEER PLAN.

BODY PLAN.

½ BREADTH PLAN.

½ DECK PLAN.

Fig. 30. Lines of a Model Service Launch Hull.

had [attained speeds of over 21 m.p.h.   These boats are about
3 ft. long, and fired with a petrol blow lamp, which utilises
a flash boiler to supply steam to a twin cylinder single acting engine.

The great pioneers of high-speed flash-steam models were Mr.
Herbert Teague and Mr. V. W. Delves-Broughton, and their patient
work some years ago undoubtedly stimulated the interest in such
craft.   The readers of the *Model Engineer* will be familiar with
the *Folly* and the *Incubus*—a fine boat, 5 ft. 6 in. long, that
exceeded a speed of 20 m.p.h., over three years ago, and for many
seasons was quite unbeaten for speed.

Before closing this review of the types of vessels usually modelled,
some reference must be made to the hydroplane, in one, at least, of
its many forms.   These unique vessels are designed to skim on

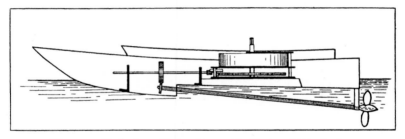

Fig. 32.   Single Step Hydroplane, with Clockwork Mechanism.

the surface of the water, and are usually flat-bottomed boats,
formed with two or more planes set at a small angle to the horizontal, and arranged with a step approximately amidships, in the
manner shown in Fig. 32.   This little boat is 21 in. long, and was
made by Mr. Hamilton for experimental work.   The motive power
is clockwork, and is so arranged with the aid of a long shaft that
the position of the motor may be varied to ascertain by experiment
the best position of the centre of gravity to secure the maximum
planing effect.   The speed attained exceeds four miles per hour,
although only for short distances, owing to the spring rapidly
running down.

A simple form of hydroplane is found in the type of vessel known
as the " Sharpie."   This is simply a flat-bottomed vessel usually
more or less triangular in plan.   Such a model is readily constructed, and gives excellent sport if fitted with light and properly

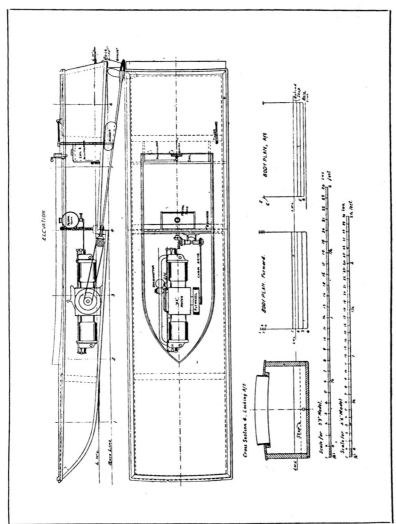

Fig. 33. Design for Petrol-engined Hydroplane.

balanced machinery. The lines of a simple hydroplane are given in Fig. 33. This shows the general arrangement of a petrol motor and the typical disposition of the machinery weights.

A modern type of hydroplane is exemplified by such a fine vessel as *Miranda IV*. This form of boat cannot easily be described. It

Fig. 34.   *Mfisto*, a Single Step Hydroplane Boat.

consists of a spoon-shaped after body, and is provided with a hard chine in the forward part of the vessel, which has a lifting effect, the vessel at full speed being supported on two points only; but if such a construction would be practical or not for model high speed work remains to be demonstrated, although the results obtained with

Fig. 36.   Mr. H. H. Groves' Record Breaker, *Irene II*.

a model of this class of boat, shown in Fig. 34, built by the author, indicate that good results are to be expected. Some further consideration of the form of such a hull is given in Chapter V on " Hull Design."

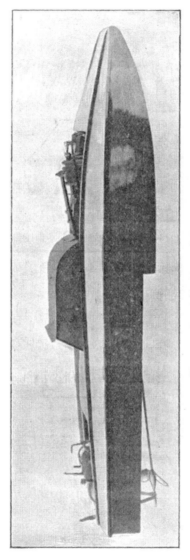

Fig. 35. Broadside View of the *Bulrush*.

Mr. Noble of the Bristol M.Y. Club has adopted a somewhat similar design but with a V floor, and this boat is shown in Fig. 35. This is the *Bulrush*, which took a Silver Medal for the splendid speed of 22.77 miles per hour in the *Model Engineer* competition for 1913. Fig. 36 shows the *Irene II*, a very wonderful little speed-boat built by Mr. H. H. Groves. Although this boat only

Fig. 37.   A Model Multi-step Hydroplane.

weighs 6 lbs. 13½ ozs. in complete running order, she accomplished the remarkable speed, for so small a boat, of 21.19 miles per hour.

The natural development of the hydroplane has taken place on two lines, one to obliterate the steps entirely, and the other to multiply them. Excellent examples can be quoted of either type to support the arguments of the respective designers, but the illustrations

Fig. 38.   Plan View of the Multi-step Hydroplane.

Figs. 37 and 38 depict a model recently finished by the North-ampton firm. She is a metre boat, Class A, has seven steps, is fitted with a Simplex engine, specially lightened with forced lubrication; a single ram pump draws the water through scoop immediately ahead of the semicircular fin amidships, and delivers it to the coil boiler, which is fired by a powerful petrol blow lamp—the details

Fig. 39. "Full Speed Ahead."

of the machinery installation being very clearly shown in the plan view.   The neat rudder gear, its position and arrangement, should be particularly noted, also the angle of the propeller shaft, which is practically horizontal.   The displacement is only 6¾ lbs. and the boat has performed well, and is quite up to the best performances of her class.

There are probably few enthusiasts who build their own model boat in which to " go afloat."   Mr. Foster, of Pangbourne, some years ago, built a fine model warship, which was used on the upper reaches of the Thames, and no doubt there are others ; but the most complete fleet of such models ever constructed was the "Fighting

Fig. 40.   Review of the "Fighting Fleet in Miniature."

Fleet in Miniature," built from designs by the author, by Bassett-Lowke, Ltd., and used for the great naval warfare display at Earl's Court during the summer of 1913.

These models averaged some 20 ft. in length, were solidly constructed of silver spruce, and carried a crew of two men.   The displacement was in the neighbourhood of 1¼ tons, as an electric motor of 1½ h.p. was employed in each, with a battery of accumulators of appropriate power.

The fleet consisted of two destroyers, two battle cruisers, four line-of-battle ships, and the Royal Yacht, *Victoria and Albert*. Fig. 39 gives a fine impression of one of the battleships, while the bulk of the fleet can be seen in Fig. 40, showing the review on the

Fig. 41. H.M.S. *Thunderer*.

River Nene at Northampton. The *Thunderer* (shown in Fig. 41) measured over 20 ft. in length, and, in common with all the fleet, was completely equipped with guns to fire, searchlights to operate properly, flags, code signalling devices, torpedo nets to work, and the hundred and one details of a typical warship, all reproduced faithfully and approximately one-thirtieth full size. The man forward controlled the speed and direction of the vessel while the man aft fired the guns, worked the torpedo nets, signals, etc., etc. The method of ingress to the vessel, and some idea of the completeness of the detail, can be gathered from Fig. 42, which shows the leading deck-fittings on the *Neptune*, one of the battleships.

Fig. 42. Some Details of the Deck Fittings on H.M.S. *Neptune*.

This fleet had six months' hard work, and was highly satisfactory, performing all its functions in an admirable manner.

Before closing this brief résumé of naval models, some reference must be made to " water line " models. These are, of course, never used as working models, but for demonstration purposes. A unique fleet of such ships, designed some years ago by the author

Fig. 43. A Fleet of Water Line Models as used by the Navy League.

for the Navy League, is shown in Fig. 43, and indicates that a
fine characteristic model can be produced on such lines. The
boat at the top of the illustration is the *Lord Nelson*, which is
divided longitudinally to show the interior arrangements, thus
materially enhancing the attraction and value of the model.

Model Electrically-driven T.B.D. *Tango*.   Built by Mr. E. W. Hoggard.
Length, 5 ft. ; beam, 6 ins. ; depth, 4½ ins.

CHAPTER III.

THEORETICAL CONSIDERATIONS AND THE QUALITIES OF A MODEL.

THERE is probably no other branch of model engineering that presents so many different, and often conflicting, aspects as naval model architecture, and the student of this subject finds, as he proceeds further and deeper into its mysteries, that the more he learns the less he apparently knows, and, consequently, the manifold difficulties to be overcome in the succesful design of a model power boat seem to be almost insuperable. The desire to provide stability, for instance, indicates the need for a broad boat with a hard bilge. The demand for speed necessitates fine lines, and a minimum of skin friction, the exact opposite of the former case ! To overcome all the difficulties and to combine them into one harmonious design is the aim of all naval model architects, and this chapter is intended to indicate the best lines on which to commence, and as far as possible show *how* and *why* the differing factors in a boat design are used. But before attempting the design of a model boat, the why and wherefore of the first principles of naval architecture should be ascertained ; then with this knowledge the student should think out for himself the relationship of every line used to indicate the form of his boat, and try to foresee as far as possible the result to be expected. Many people suppose that model boating is only a childish pastime ; be that as it may, the fact remains that a study of the scientific side of the subject, and a participation in its sporting aspects will show that it is more than this, and is indeed a most fascinating scientific study.

This book can only hope to indicate one door through which the subject may be reached, that is, by appreciating something of the laws governing model boats, which to a large extent also govern the design of a mammoth liner or a tiny row boat.

The first of many phenomena to be considered arises from the power a ship possesses, or should do, to float upon the surface of the water, and to maintain proper trim and stability, that is to say, to float upright and to her designed water line. At the first glance it seems impossible that a model constructed largely of metal and materials which are in themselves considerably heavier than water should possess the power of flotation. This is due to what is known as "displacement" and "buoyancy," that is to say, when a vessel or boat is floating in water it displaces, or pushes away, a volume of water having a. weight equal to the total weight of the boat. To make the matter clearer—suppose we have a bowl of water containing to the brim exactly six pounds weight of water, and into this we place a rectangular block of wood weighing exactly one pound, having first placed this bowl on a large plate or dish. When we put the wooden block into the water a certain quantity will overflow and run down into the dish. It we remove the bowl with the piece of wood, and then weigh the amount of water remaining in the dish, we shall find it weighs exactly one pound, which is the weight of the wooden block. If we had put a solid iron weight of one pound into the bowl of water, the result would have been quite different, for the iron would sink to the bottom, as it is the *volume* of displacement which enables a vessel to float. The first point therefore to bear in mind when designing a vessel is that the amount or *volume* of that part of the vessel actually immersed in the water displaces a weight of water exactly equal to the total weight of the boat. This principle is the most vital of the many that have to be borne in mind in designing a model boat. It is essential, for a body to be able to float, that its volume immersed, or the part of it that sinks *into* the water, shall displace or push aside an amount, or volume, of water equal in weight to the total weight of the body. Thus, if a boat weighs 10 lbs., it must displace 270 cubic in. of water, and still leave some portion of the hull above water ; because 27 cubic ins. of fresh water weigh one pound, therefore ten times 27 cub. in. equals 270 cub. in. which represents the weight of the boat, that is, 10 lbs.

The diagram Fig. 44 shows a boat floating in a bath of water; the immersed portion of the hull, A, displaces 270 cub. in. of water. That is, the "volume of displacement" is 270 cub. in., the weight of the boat is 10 lbs. The weight of the 270 cub. in. of water being 10 lbs. the boat is able to float. It must be noted here that this volume of the boat is that part which is immersed, or *in* the water; an extra amount must be provided to form the " freeboard," or *above water* portion of the hull on which to erect the bridge, funnels, etc. But even after we have constructed a vessel which will float, it is necessary for the success of the model that it floats in an upright position, and moreover that it is not unduly liable to capsize ; in other words, the vessel must possess stability. Stability may be defined as the property a vessel possesses to return to her upright position after having been slightly inclined .from it. These inclining movements are of two classes—Side Rolling, or Heeling, and Longitudinal Rolling, which is known as alteration of trim.

Fig. 44. Boat Floating in Bath of Water.

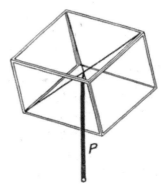

Fig. 45. Illustrating position of Centre of Gravity.

There are two conditions of equilibrium of a floating body :—

1. The weight of the body must equal the weight of the water displaced.
2. The centre of gravity and centre of buoyancy must be both in the same vertical line.

These two factors will insure the boat floating but not necessarily in an upright position. For a vessel to float in the desired upright position a third condition is essential.

3. The metacentric height must be positive, that is to say, it must be above the centre of gravity.

It is most difficult to describe in letterpress the fundamental laws of stability, but the author hopes that a little further explanation will tend to elucidate some of these problems.

Now, when a body possesses bulk and mass, that body has weight, and there is somewhere in that body a mean centre, upon which that body could be balanced were it accessible. That balancing point is called the *centre of gravity*, because the force of gravity always acts vertically downwards through that centre, no matter how much the position of the whole mass may be changed. The diagram Fig. 45 is supposed to represent a rectangular box, and if diagonal struts of no magnitude were put across from corner to corner as shown, and a pointed rod P placed in the exact centre of the mass, the whole arrangement could be balanced on that spot, no matter how much the box was moved ; that spot is called the centre of gravity : it is assumed, of course, that an " invisible rod " could be used !   As already stated, the centre of gravity *never alters* in a boat unless the weights in the boat are themselves altered, and the force of gravity *always acts vertically downwards through the centre of gravity.*

Buoyancy is virtually the reaction due to gravity, and every portion of a vessel at rest and floating in water may be considered as experiencing a pressure perpendicular to each portion of the surface immersed, and the sum of these pressures is termed buoyancy, and must equal the weight of the vessel.

The centre of buoyancy is the centre of gravity of the immersed volume of the model, and is the point at which the buoyancy is assumed as concentrated.   When the model is upright, the centre of buoyancy is in the longitudinal middle line of the model, but when inclined or heeled, the centre of buoyancy moves towards that side of the ship which is depressed.

The calculations for ascertaining the position of the centres of gravity and of buoyancy are determined by an application and extension of Simpson's Rule, and are described in Chapter V.

The righting power of a vessel, or the capability to resist heeling, depends almost entirely upon the design or form of the hull, the ratio of depth to beam, and the relative heights of the centre of gravity and centre of buoyancy.   In general parlance, this faculty is known as " stability."   We will first of all consider the case of Transverse Stability, or the power to resist a heeling force.

When a vessel floats in an upright position, the centres of gravity and buoyancy are both in the longitudinal centre line, and vertically under one another.   This will be explained best by the aid of a

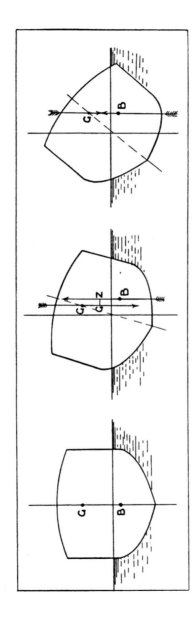

Fig. 46.   Fig. 47.   Fig. 48.

Illustrating Centre of Gravity in relation to Centre of Buoyancy.

few diagrams. Fig. 46 shows a cross section of a vessel immersed in water. The centre of gravity is indicated at G, the centre of buoyancy being situated at B, both forces acting in equal and opposite directions. The vessel floats in a normal vertical position ; but some horizontal force causes the vessel to incline to the right and assume a position shown in Fig. 47. The centre of gravity of the vessel will still remain unchanged, and still remains at the same spot, G, on the centre line of the boat, but the centre of buoyancy will move over to leeward. We now have two forces, one due to the weight of the vessel, and acting downwards through G, and another, also equal to the weight of the vessel, but acting upwards through the centre of buoyancy and due to the flotational powers of the vessel. Two such equal, opposite, and parallel forces are termed a couple, and the moment of the couple is the product of either force into the perpendicular distance between them. Thus the length G Z, multiplied by the weight of the boat, is equal to the moment of " Statical Stability," the length G Z being the perpendicular distance between the lines of the two forces, and being termed the stability lever. By inspection of Fig. 47 we observe that the centre of buoyancy acting upwards and to leeward of the centre of gravity, through the agency of G Z (the stability lever) exerts a righting motion to the vessel, causing her to resume her upright position ; in other words, the vessel is stable, the buoyancy of the vessel acting somewhat in the manner of a man lifting a weight with a lever. Suppose, upon the other hand, that the vessel be inclined to a much greater angle, as indicated in Fig. 48 It will be observed that the centre of gravity and centre of buoyancy both coincide, in which case the vessel would remain in this position, there being no righting movement whatever. Should the vessel, however, be heeled still further, the centre of buoyancy will act to windward of the centre of gravity, and accelerate the heeling movement, causing the vessel to capsize. It has been found by experiment that through small angles of inclination the line of action of buoyancy cuts the vertical line, or line of buoyancy when the vessel is in an upright position, at a fixed point M, and known as the Metacentre. The relative position of the centre of gravity and M, the metacentre, determines the stability of the vessel. When M is above G, as in Fig. 49 at A, the vessel is in a stable condition. When M and G coincide, that is to say, when the metacentric height is nil, as at B, the vessel will remain in any position to which she may be forced, a state of affairs known as neutral equilibrium. When, however, M, the metacentre, is below G, as at C, Fig. 49, the forces will tend to make the vessel heel still further from the upright position ; in other words, the vessel will be in unstable equilibrium, and turn over. The height of the metacentre depends solely upon the immersed form of the vessel, and is quite independent of the length of the ship. The height is determined by first finding the vertical position of the centre of buoyancy, and then the distance between B and M, Fig. 49. This distance B M is equal to the moment of inertia of water plane area about the middle line divided by the volume of displacement, and is strictly proportional to the square of the breadth, and inversely proportional to the depth. In the

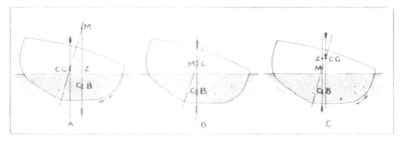

Fig. 49. Righting Moments and Metacentric Heights.

case of a vessel with a high metacentre, she will be what is known as stiff, and will not roll considerably. With a medium or small metacentric height, we shall have a boat with a low period of oscillation, due to the fact that the centre of buoyancy will not be driven far from the normal centre line of the vessel, and, consequently, such powerful forces will not be brought into play. It is often thought that the application of ballast to the lower parts of a vessel will cause her to be steady, but this is a mistaken idea. The vessel will certainly be stiffened, but she will roll badly ; an even distribution of the necessary weight to insure displacement is a far more satisfactory method of obtaining a stable and useful model. The real value of these facts will be better appreciated as we go further into the question of the design of a model vessel.

In considering the problem of the stability of models, it is essential to remember that the shape and volume of the displaced water, wherein the vessel is floating, determines the position of the centre of buoyancy, and hence the stability or otherwise of the boat, as the movement of the centre of buoyancy from its original position will continue until it again finds the mean centre of the displaced water, which centre depends upon the shape and volume of the water so displaced.   According to the position of this centre relative to the centre of gravity, the stability or otherwise of the model depends.   If the centre of buoyancy moves outwards *beyond* the centre of gravity, there will be a righting force or leverage tending to push the boat upright.

This is, however, only part of the tale, as there yet remains the question of longitudinal stability, or power to resist changes of trim.   This is really only another aspect of the same laws, but is usually dealt with separately, and it is hoped to show, with the aid of another diagram, how these factors can best be dealt with ; as who has not made a model, only to find that when put in the water she is so much down by the head that the propeller is hardly in the water at all, or that the stern is practically sunk below the waves, while the bows are high and dry !   It is quite easy to avoid such a contretemps, however—when you know how—and to arrange it all before the boat is made or even the final lines of the hull are drawn.

With a vessel floating at rest, the *longitudinal* position of the centre of buoyancy must come directly under the centre of gravity, for the same reasons already explained in the consideration of transverse alterations or heeling.   Consequently, if the position of the centre of gravity is altered by moving the position of the engine, boiler, etc., forwards or aft, the centre of buoyancy will have to shift in like manner, until it again comes under the centre of gravity (at least at ordinary speeds).   Therefore, the correct placing of the weights in the boat is one calling for especial care, as it affects considerably the lifting power or otherwise of the model.   It should be apparent that when the position of centre of gravity alters, the boat must sink until the immersed portion of hull forwards or aft of the normal centre of buoyancy displaces an amount of water whose weight equals the weight of the complete model forwards or aft of the centre of gravity.   This feature applies perhaps more to sailing models than power boats, except in so far as it has a great bearing on the action of a boat at speed, and shows the necessity for correct placing of weights, and proper proportioning of the forwards and after bodies, as the portion ahead of and astern of the midship section, or centre of gravity, is termed. Thus, to prevent " pitching," or sinking by the head, a good flare is needed at the bows, and, as far as possible, a *reserve* of buoyancy aft to counteract the evil of " squatting," or sinking of the stern.

The proportions of beam to length and depth to beam, greatly affect stability. Thus a long, narrow boat with a fairly deep body is usually, though not necessarily, unstable. A broad, shallow boat, on the other hand, is usually very stable, until considerably heeled ; hence we have the various general *types* of hull, the broad shallow racer, the rounded body of the lifeboat to ensure an easy sea-boat, the rectangular liner to give maximum carrying capacity and so on ; but before going further with our subject, attention should be given to the relative stability of these various characteristic types.

The open " sharpie," or box form boat, has the maximum initial stability, and greatest stability at no angle of heel ; but it quickly falls when the boat is considerably inclined. The circular section, on the other hand, possesses a very low stability, but is easier in a sea way and has the lowest wetted area and least skin friction. A compromise, therefore, is the general result of attempting to reconcile these differing factors, and the well-known T.B.D. section is probably still the best fast all-round section for a large or small boat.

Fig. 50 shows four characteristic midship sections.

A. The box form of sharpie. This form gives the maximum of stability, when on an even keel.

B. A well-rounded section, such as may be used for small steam pinnaces, or lifeboats. The rounded sections have the minimum skin friction, but stability is

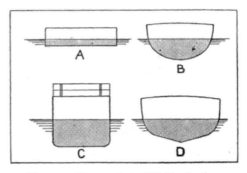

Fig. 50. Characteristic Midship Sections.

not good. These boats roll badly, due to a comparatively short "righting arm."

C. The rectangular or box form, as used on liners, cargo boats, and such merchantmen generally. One of the reasons for this section is to obtain the maximum of cargo space.

D. The T.B.D. section, a compromise between the foregoing. The beam is reduced as far as possible to keep the lines fine ; the shallow draft allows of a reasonable beam draft ratio; the rise of the floor, and easy bilge, reduces wetted surface, and tends to an easier sea-boat.

Before going further, it would perhaps be as well to define some of the centres and coefficients which must be used in preparing the design. We have already dealt with displacement and stability. The centre of buoyancy may be defined as the centre of gravity of the immersed portion of the boat, that is to say, the centre of buoyancy is the exact centre of the mass of the vessel *immersed in the water.* The centre of gravity is the centre of the *whole* mass of the vessel. In the case of models of ordinary constructions, these centres may be taken as being in the central vertical plane of the vessel, and it only remains to ascertain their fore and aft position. These positions are found by an extension of Simpson's Rule. The method of obtaining the fore and aft position of the centre of gravity of a model already made may readily be found by simply finding that position of the boat where it may be supported without falling forwards or aft. In the case of small or moderate sized vessels this can easily be done with the aid of the two forefingers, or by balancing the boat on a narrow strip of wood ; but when we are designing a new vessel, it is only possible to find these by calculation, and the exact position of the centre of gravity is found by taking moments about two planes ; the planes usually selected being the load water line and the midship section. The load water line is used to ascertain the vertical position and the midship section is taken to ascertain the fore and aft position. To calculate the centre of gravity all the weights of the vessel are multiplied by their per-pendicular distances, from one of these planes, and the products added and divided by the total weight, the result giving the distance of the centre of gravity from the planes. One set of calculations are, of course, carried out relatively to the load water line, to ascer-tain the vertical position, and another set are carried out to find the position of the centre of gravity fore and aft. To find the fore and aft position of the centre of buoyancy further reference must be made to Simpson's Rule, which is gone into at some length further, the method being to multiply the " product for volume " of the plane by the number of intervals, add the products, and multiply by the common interval and divide by the sum of the products for volume, the result giving the distance of the centre of buoyancy from the planes.

There are three principal coefficients, of which note will have to be taken in designing a model boat, the first being the *Block coefficient* or coefficient of fineness. This may be defined as the ratio the volume of the underwater body bears to a rectangular block, whose length is equal to the load water line, the breadth equal to the beam, and depth equal to the mean draft of the model in question. The cubic area of the underwater part of the boat divided into the cubic area of the rectoid gives the block coefficient. For this purpose, of course, the load water line plane is taken as being the top of the block. The average block coefficient for model battleships is .6 to .65; for cruisers .5 to .55; model T.B.D's .45; for model racing boats. of extreme type .35 to .45. Cargo boats and other bluff boats have a coefficient as high as .7 to .8 The *Prismatic coefficient* is the ratio the volume of the underwater body bears to the immersed area of the midship section multiplied by the length of the water line. The use of this prismatic coefficient is extremely valuable in the early stages of model boat designing, as if we know the weight required in pounds and the length of the load water line, we can find the suitable area of immersed midship section, the method being extremely simple. Merely divide the displacement in cubic inches by the length on the load water line in inches, and the result by the prismatic coefficient. The third coefficient to which particular attention will have to be directed is the *Midship coefficient*, which is the ratio of midship section area to a rectangle having the same beam and draft as its dimensions. The ratio may vary from .65 to 1, which would be a plain rectangle, the average being .75 to .80.

TABLE No. 1.

TYPICAL COEFFICIENTS.

BLOCK COEFFICIENTS.

Tug Boat	..	..	..	.42
T.B.D.	..	..	..	.40
Cruiser	..	..	..	.51
Motor Boat ..		..	..	.48

PRISMATIC COEFFICIENTS.

Tug Boat	..	..	..	.45
T.B.D.	..	..	..	.55
Cruiser	..	..	..	.71
Motor Boat ..		..	..	.84

MIDSHIP COEFFICIENTS.

Tug Boat	..	..	..	.80
T.B.D.	..	..	..	.75
Cruiser	..	..	..	.88
Motor Boat ..		..	..	.90

There is one underlying principle of design practically invariably adopted for the most satisfactory and successful results, and this is based upon what is known as the Wave Form Theory. This wave form theory is very properly founded upon observation and experiment, and shows that the cross sectional areas of the underwater fore-body, when plotted in the form of a curve, should approximate to a curve of versed sines, and that the area of the underwater after-body should approximate to a curve of trochoid. It must be mentioned that many designers do not consider this theory of much weight, but experiments show that it is the form of the " curve of sectional areas " that determines the speed of a vessel, and the wave form theory certainly does give a very good basis for this. It will perhaps be more explicit to take an actual example, and work through all the processes of design, and for this purpose the author has selected the design of a full sized metre torpedo boat destroyer of the popular "river class," and every process is fully detailed in Chapter V; but before commencing a description of the actual method of designing, it is necessary to consider the question of the resistance of the model to propulsion through the water.

Medal presented to the Author by the Modéle Yacht Club de Paris

CHAPTER IV.

RESISTANCE AND PROPULSION.

THE consideration of resistance, or the force that tends to stop the passage of a model through water, has always been one of the greatest interest to the naval model architect, also to the practical boat builder and user. It is obvious that the resistance to be overcome at a given speed determines the power required for a boat ; but as in model work the machinery is always as large as possible, the problem then becomes how to reduce the resistance of a given design or boat, so that the maximum speed of travel may be attained from the power of the engines. The resistance of a model to its passage through water is due almost entirely to five factors, and does not necessarily depend merely upon the actual weight of a boat, or to the actual fineness of her lines, but to a combination of all these, and other considerations as well. The chief causes of resistance in model power boat work are the following :—

1. Skin Friction, which is determined by the extent and nature of the wetted surface.

2. Eddymaking, a disturbance of the water caused by partial aeration and internal movement in the water.

3. Wavemaking, which is almost entirely governed by the formation of the hull and the fineness or otherwise of the lines.

4. The Wake, or sternward wavemaking, flowing in the same direction as the boat, and due to a combination of all the other causes.

5. Air Resistance, caused by the air pressure, governed by the relative surface areas, and the speeds of the boat and the wind.

Frictional or *Skin Resistance* is caused by the friction of the water on the immersed surface of the ship. The laws governing it have been determined principally by the experimental researches of the late Dr. W. Froude. He found that the resistance varies with the nature of the surface, with the length of the surface in the direction of its motion, and approximately with the square of the speed. Dr. Froude's experiments were made with thin planks of various lengths towed through water ; the planks were fined at the extremities to get rid of eddy resistance, and suspended from the carriage of the experimental tank, so that they moved endwise with their whole surface submerged. The surfaces were in turn covered with tinfoil, paraffin wax, varnish, fine sand, medium sand, and coarse sand. With a length of 2 ft. the resistance in pounds per square foot, at 10 ft. per second, varied from .3 to 1.1 lbs. for the various surfaces. The resistance of varnished surfaces, 2, 6, 20 and 50 ft. long, at 10 ft. per second, were respectively .41, .325, .278 and .250 lbs. per square foot, and for different speeds the resistance of these lengths of varnished surfaces varied as the speed respectively to the 2.0, 1.85, .185, and 1.83 powers. The reduction in resistance per square foot as the length is increased is caused by the forward part of the plank imparting to the neighbouring water a velocity in the direction of motion, thus causing the after portion to pass through water moving in the same direction as itself. The water caused to move in the same direction as the plank is termed the " frictional wake," and is of some importance in connection with the action of screw propellers. As skin friction is caused by the roughness of the hull when travelling through the water, and cannot be entirely eliminated, it must be reduced by making the hull as smooth and fair as possible, and also by such expedients as the use of lubricants, *e.g.* black lead, or heavy oil smeared on. But skin friction, apart from being a direct cause of loss of power, produces other effects, such as that of the wake, and the " area of frictional disturbance " in the water due to distortion of the stream lines of the water in the vicinity of the hull. The experiments conducted by the late Dr. Froude have enabled the skin friction to be accurately determined, and these results are briefly as follows :—

Resistance per square foot of wetted surface taken as a mean over the whole surface =

Varnish	..	..	..	.41
Paraffin Wax		..	..	.38
Tinfoil	..	..	..	.30
Coarse Sand ..		..	..	1.10

These results were obtained at a speed of 10 ft. per second = 5.92 knots = 6.8 m.p.h.

The actual resistance of an average model is about .25 lbs. per square foot of wetted surface at a speed of 6 m.p.h., and in a metre model of average form would be in the neighbourhood of $\frac{1}{2}$ lb., assuming a surface of 280 square inches immersed, at a speed of approximately 7 m.p.h. (6.91 m.p.h.).

But of course this varies with the speed variation, and also varies as the 1.83 power of the speed. The calculation is somewhat intricate, and the table of values, Table No. 2 (Prof. J. H. Biles), should be of help to the designer.

The formula for the calculation of frictional resistance due to skin friction is—

$$RF = FM \times M \times S$$

RF = Frictional resistance in pounds.

FM = Mean resistance in pounds per square foot of wetted surface.

$$M = \left\{ \frac{V}{5.92} \right\} 1.83$$

S = Surface of immersed hull in square feet.

TABLE No. 2.

PROFESSOR J. H. BILES.

FRICTIONAL RESISTANCE.		
V = Speed.	$M = \left\{ \frac{V}{5.92} \right\} 1.83$	EXAMPLE.
1 —	.0386	A model boat has 2 sq. ft. of
3 —	.2883	wetted surface.
5 —	.7341	FM = .41 lb. per sq. ft.
7 —	1.349	= .82 lbs. at a speed of
9 —	2.152	6 m.p.h.
11 —	3.107	To find resistance at 11 m.p.h.
13 —	4.219	Under column V = Speed
15 —	5.482	(Table No. 2), at 11 read in
17 —	6.893	column M. 3.107.
19 —	8.448	∴ substituting values in the
21 —	10.146	Formula.
23 —	11.985	RF = .82 lbs. × 3.107.
25 —	13.961	= 2.54774 lbs.
27 —	16.071	or 2$\frac{1}{2}$ lbs. nearly, at a speed of
30 —	19.489	11 m.p.h.

The above should make clear the enormous increase in resistance as the speed increases.

There have been from time to time many formulæ put forward for calculating the total resistance of a boat, but as these are all based on previous trials and assumed hull forms and types, they are of little value in model work, although when used with discretion, and some amount of practical experience, give good results. Of course, their value to the naval architect, when calculating the design of a full sized vessel of normal type, is very great and they are usually quite reliable.

The author has found that as a rough guide to model resistances it is reasonable to take the skin resistance at the desired speed, and add 25 per cent. to cover other losses, due to wave-making, &c. Thus, if a boat at, say, 10 m.p.h. has a skin resistance of 4 lbs., the addition of an extra 1 lb., making a total resistance of 5 lbs., will be about right for most boats of ordinary normal type. It is probable, however, that the resistances are rather less than more in small models.

Eddy-making is caused by the presence of sharp angles, sudden changes of form, and appendages, such as shaft brackets, rudder posts, etc., and is sometimes a fruitful cause of loss of power, although the model boat of modern design is not such a bad offender in this respect as the earlier models. Eddy-making is often to be observed at the stern of flat transom boats, when the speed is such that the water is unable to flow in behind them fast enough. This effect often accounts for the difficulty in steering such a model at speed. In a well formed ship the eddy resistance is small ; it is caused by the formation of eddies·in the dead water behind the blunt ends of the after portions of the ship, just as at the back of a flat plate towed through the water.

Wave-making as affecting the resistance of a boat is a subject of entrancing interest to the naval architect, and necessarily so to the model boat builder, as the reduction of wave-making means that a greater power from a particular engine is available for speed. Let us first consider what is a wave. Lord Kelvin defines it as " the progression through matter of a state of motion," or as the " progression of a displacement." This is a noble and scientific statement, of which it is easy to demonstrate the truth by a simple experiment with a piece of cord. Tie one end firmly to a post or

F

solid fixed object and, grasping the other firmly, shake the hand up and down, when you will see the wave action progress along the cord, and also note by the relative rise and fall of the hand the *height* from crest to hollow of the wave, and the spacing or number of waves per unit of length.

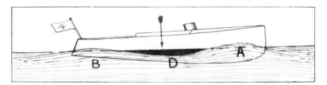

Fig. 51. Wave-making Resistance.

When a model or a ship is at rest, the pressures due to gravity are practically evenly distributed over the whole surface of the hull, but when the ship is put into motion, the " still water and at rest " distribution of forces is altered with a corresponding alteration or increase of pressure, in one part, with a diminution of pressure in

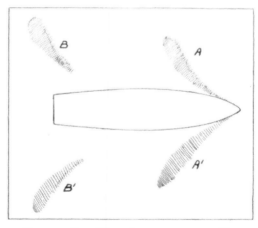

Fig. 52. Bow Wave System as seen in Plan.

others ; but the *surface* pressure due to the weight of the atmosphere always remains constant, hence differences in level are apparent, otherwise waves are produced.

When a boat passes through a medium such as water, the normal

arrangement of the stream lines are upset, their distortion at the
bows, as in diagram Fig. 51, causing their speed to increase at A with
a corresponding decrease in pressure, hence a bow wave system is
formed.   Again, at the stern, we have a similar state of affairs,
the decreasing speed of the stream lines at D causing increase of
pressure, and the formation of the stern wave system at B.

Taking the bow wave system first, this, of course, is the result of
the passage of the forepart of the vessel through the water, and is a
regular wave of translation, that is to say, the energy put into the
water to cause and maintain the wave system is entirely lost, as
these waves pass right away from the ship, and no part of their

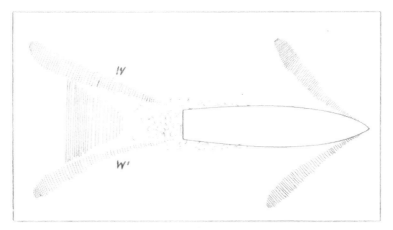

Fig. 53.   Stern Wave System and Wake.

energy system is returned to the vessel.   In Fig. 52 such waves
are shown at A A¹ and B B¹.

The middle portion of a vessel, which is usually more or less parallel
in large vessels, but very seldom so in models, causes little or no extra
wave-making resistance, even when this middle body is practically one-third the entire length of the ship ; the only practical
resistance of this portion at ordinary speeds being due to skin
friction. This accounts in a large measure for the *increased* efficiency
of large vessels over properly proportioned smaller ships.

The actual state of affairs in the region of the after body of a vessel is still largely a matter of conjecture, but the run or after body requires the very greatest care in design to secure satisfactory results, because it is certain that another and independent system of waves, W W¹ Fig. 53, is formed by the passage of the after portion of the vessel, the natural tendency of the water being to flow into the cavity which would otherwise be left in the water when the vessel had passed, and thus the diverging stern wave system is generated. Then we must remember that the mere skin friction of the hull on the water causes a forward movement of a belt of water adjacent to the hull, and known as the " zone of frictional disturbance."

The limits of this work prevent a full exposition of the theory of wave-making, and the chief causes thereof, but the latest results of experimental work indicate that the character of the " curve of sectional areas," described fully in Chapter V, largely determines the resistance of the hull, and governs the formation of the wave systems.

The varying speeds of a model also affect the wave-making resistance. At low speeds, of 2 or 3 m.p.h., wave-making with a properly designed boat, is practically non-existent, but as the speed increases, the wave-making resistance increases, until a maximum is reached, when it falls off somewhat, although a further increase is noticed with additional increases of speed.   This phenomenon is due to several causes, one being the interference of the bow and stern wave systems, another factor being the relative resistances of the surface (skin friction), at varying speeds ; while alterations of trim such as a lifting of the bows, contribute to this. It can only be calculated for a model when the lines have been prepared, and then the process is intricate.

Wake Resistance. This resistance, caused partly by eddy-making, but chiefly by skin friction, is due to the passage of the vessel having set in motion in a forwards direction, the surface and immediately surrounding water; resulting in a loss of energy and speed ; but some of this lost energy may be regained by suitable propeller arrangement, as will be seen later by a further consideration of propulsion.

Air Resistances have been generally disregarded in model boat work, but a little thought should show the serious side of this neglect, as it should be remembered, in model speed boat work, that there is a *relatively* large portion of the boat hull exposed to air currents, and air resistance should be kept in mind, as although it is only .005 A V² (A = area exposed in square feet; V = speed in feet per minute), the *added* effect due to a head or side wind is very great; for example, a model steaming at 6 m.p.h. in a calm, with say 3 ft. of surface exposed, has an air resistance of only .540 lbs., whereas with a head or beam wind, of only say 4 m.p.h., the air resistance rises to 1.5 lbs., showing how much the air resistance and air currents can affect a model, especially one with little lateral resistance.

The viscosity of the water is also, with model speed boats, a source of resistance. The precise nature and effect of this is but little known, but the subject is under investigation at the present time.

It should be noted in passing that some parallel middle body in a boat gives steadiness of running, also the position of the midship section affects speed materially, usually the higher the speed, the further aft the M S section should be; in fact, trials of hulls with equal displacements show as much as 10 per cent. increase in efficiency as regards speed by setting the M S section 10 per cent. further aft.

There is also another curious cause of resistance, at speed, viz., that due to the interference of the bow and stern wave systems. The resistance being at maximum when the crests of the bow waves coincide with the crests of stern waves, and at minimum when the crests of the bow waves coincide with the troughs of the stern wave system.

Some of the greatest problems in connection with resistance are those presented in ascertaining the power or thrust necessary to overcome the resistance, and it must be stated right away that for model work, at high speeds, the subject is shrouded in mystery. The answer is fairly easy to give if we know the total resistance (at the required speed) of the hull, with the correct weights placed in their appropriate positions. This, in practice, is usually found by " towing the model hull with a brick or two in it ! " Of course, the results so obtained are quite valueless for all scientific purposes, as the model will behave beautifully while the tow rope is on it, but afterwards—well—that's another tale.

The tow rope pull, however, provided the distribution of weights in the hull follow the designed arrangement of the proposed machinery, gives us the only *practical* guide to the power needed when the thrust required (or pull exerted on the rope) is known at the desired speed ; the method may be approximately calculated as follows :—

$$\text{Formula.} \quad \frac{RV}{33,000} = \text{E.H.P.}$$

In the formula :   R = Resistance in pounds,
V = Speed of boat in feet per minute,
E.H.P. = Effective power at the propeller,
and the *total* power required is this amount plus that consumed in engine and shaft friction, etc., usually 40 to 60 per cent. in models.

To ascertain the total resistance of a model, two courses are open : (a) To tow the actual hull at the desired speed, and record the resistance in pounds by means of an accurate spring balance ; (b) To approximately judge the resistance is to calculate the skin friction as already explained, multiply the result by the speed of the boat in feet per minute (see Table 6, Chap. V) and divide by 33,000 to ascertain the horse-power of the energy needed, as 33,000 ft. lbs. represent one horse-power of mechanical energy. As, however, for small boat work the result would come out generally in the form of an odd fraction, it is best to calculate the horse-power in Watts or the measure of electrical energy. On the assumption of 750 Watts equals 33,000 ft. lbs. the formula becomes

$$\text{E.H.P. in Watts} = \frac{RV}{44.}$$

because 44 ft. lbs. mechanical equals 1 Watt electrical measure. For example, a boat has, say, 2 lbs. resistance at a speed of 500 ft. per minute.

$$\text{Watts} = \frac{2 \times 500}{44} = 23 \text{ (nearly)}.$$

That is, 23 Watts is the electrical measure of the total resistance of the model, and assuming an all-round mechanical efficiency in the engine, shafting, and propellers, of 50 per cent., the engine would have to give an output of 46 Watts to drive the boat, 23 Watts being used up in frictional and other losses in the engine and transmission, leaving 23 Watts available to drive the boat. Had this been calculated on the horse-power basis the nearest equivalent would be

.032 H.P. The advantage of using "Watts" as the measure is apparent, and of course would be required in the case of electric boats.

The author, during the autumn of 1911, made some total resistance tests with various hulls, and the results are tabulated in Table 3.

They were made on a fairly calm day, on the River Thames, the method being to tow the models, as suggested above in paragraph (a), from a long beam extended from the side of a fast motor boat, kindly placed at the author's disposal by Mr. W. J. Tennant. The actual speed of the motor boat was ascertained by a special Walker's Log, and the tow rope pull, or total resistance, taken by means of a delicate spring balance, a fine cord, running over practically frictionless pulleys, being used. The results given are the mean of several runs, but must be taken with caution, as they only represent the naked hull resistances, no shafts, rudders or other appendages being fitted ; but will serve as a basis in selecting a suitable propeller, etc.

Messrs. Bassett-Lowke, Ltd., kindly placed a number of different hulls at the author's disposal, and they were all trimmed with lead ballast, arranged to represent the machinery weights as accurately as possible.

A brief consideration will show that with boats of the ordinary type a limit of speed is rapidly reached, beyond which it is not practicable to drive the hull, while if an excessive power is provided, the hull will lift, and an apparently impossible state of affairs exists. The volume of water displaced at any given instant of time no longer represents the actual weight of the boat. There are several theories to account for this, but the time factor, in conjunction with the viscosity of the water, appears the more probable ; just as the density of the air supports the moving planes of an aeroplane, so the density or viscosity of the water supports the boat. The time during which the water could support such a weight being very limited, it follows that if the speed of the boat falls relatively to the " critical speed " of the boat, the foregoing laws of displacement all hold good, but if the speed increases very greatly, the boat, if of suitable design, will skate or slide *over the surface* of the water, and not to any great extent force its way *through* it. Such vessels are, of course, always very highly powered relatively to their displacement, and are —as a class—known as hydroplanes.

MODEL POWER BOATS

TABLE No. 3.

RESULTS OF RESISTANCE EXPERIMENTS.

BY E W. HOBBS.

Tests made on River Thames at Putney, November, 1911.

Hull.	Displace-ment.	Speed.	Mean approx. total Resist-ance.	
	lbs.	m.p.h.	lbs.	ozs.
5 ft. 6 in. Torpedo Boat Destroyer	24	8	2	
		6	1	8
4 ft. 6 in. do. do.	18	6	1	4
Metre do. do.	10	8	1	4
do. do. do.	10	6	1	2
do. do. do.	10	4		12
Tugboat (metre)	14	4		9
24 in. Hydroplane. (Very dry boat.)	4	8		10
do. do. do.	4	6		8
do. do. do.	4	4		4
30 in. Racing Hull. (Sank quickly, limit of speed, 5 m.p.h.)	4	4		8
36 in. Cargo Boat. (Long easy lines, very steady running.)	6	4		7
do. do. do.	6	6		12
30 in. Vedette Hull. (Round easy sections ; a good sea boat.)	5	4		9
do. do. do.	5	6	1	4

The *total* resistance of a model power boat, being ascertained, the problem then presented is how to overcome that resistance and drive the vessel at the desired speed ; in other words, what system of propulsion shall be adopted.

There are available various systems, such as the Jet propeller, paddle wheels, oscillating vanes, and other devices, but it is possible that aerial propellers, by reason of the freedom from appendages to the hull, will find favour in future with model racing enthusiasts, and provided the hull has ample beam and a well designed screw is used, quite excellent results should be obtained. So far, however, of all the various methods of propulsion available for marine use, the screw propeller of submerged type ranks highest because it is at present the most efficient and practical. One of the earliest recorded screw propellers dates back to 1752, and was invented by De Bernonvilli, who mentioned a submerged propeller with blades at sixty degrees.

A screw propeller is defined by Professor Biles as "An apparatus which by its rotation about a straight line produces motion in the direction of that line."

The ordinary screw propeller usually has two, three, or four blades, identically alike, equally spaced, and more or less of a helical

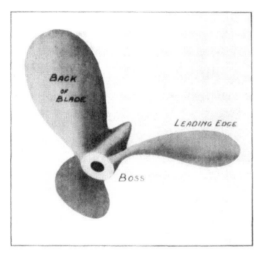

Fig. 54.   Typical Three-bladed Screw Propeller
with Conical Boss.

surface. The blades are mounted on a central boss, which is provided with means to secure it to the shaft, an excellent example being illustrated in Fig. 54.

A *Right-handed* propeller is one which, when viewed from aft, turns in the direction of the hands of a clock, and drives the vessel forwards.

A *Left-handed* propeller has the blades at an angle opposite to that of the blades of a right-handed propeller, and revolves anti-clockwise when driving the vessel forwards.

The *Driving face* is the after side, and this surface is used to drive the vessel forwards.

The *Back* of the *blade* is the forward side; its shape depends upon the design.

The *Leading edge* is that portion of the blade which first cuts the water when turning ahead.

The *Disc circle* is the circle struck out by the outer tips of blades when revolving.

The *Disc area* is the area of this circle.

The *Diameter* of a propeller is the diameter of the disc circle.

The *Pitch* of a propeller, as a whole, is the linear distance that the propeller would travel in one revolution, if its driving face worked in a solid of the same form, as in the case of a bolt turning in a solid nut.

The *Pitch* of a propeller is not uniform, and as it may vary axially or radially, an average has to be taken in both cases.

Radial pitch variation is ascertained by considering the variation of pitch at different points of a radius.

Axial pitch variation is shown by considering the variation of pitch on a circumference of a circle drawn on the blade, with the axis as centre.

Area of blade is the developed area of the driving face, and is the helicoidal area.

Area of a *propeller* is the sum of the areas of all its blades.

The *Projected area* is the projection of the helicoidal area on a plane perpendicular to the axis.

$$Pitch\ ratio\ = \frac{P}{D}$$

That is, *pitch* in *inches* divided by diameter in inches.

$$Diameter\ ratio\ = \frac{D}{P}$$

That is, diameter divided by pitch.

Slip ratio—

$\quad$ N = Revolutions per second.

$\quad$ P = Pitch in inches.

$\quad$ V = Speed of model in inch secs.

That is, the actual speed of propeller relatively to still water ; but the propeller speed or travel if working in a solid fixed nut = P N,

Slip therefore would be P N—V

$$Slip\ ratio\ =\ \frac{P\ N—V}{P\ N}$$

$$\therefore\ P\ N\ (1—\text{slip ratio})\ =\ V.$$

Action of propeller—The action of a screw propeller is to direct a column or jet of water sternwards, Fig. 55 ; of course, when submerged, the propulsive effect on a body moving through a medium such as water being due to the projection of a column or stream of the medium in a sternwards direction. In a perfect fluid this would set up a stream line motion forming a closed kinetic chain and would produce no thrust, so long as the stream line system remained unbroken ; consequently, to produce a thrust there must be a break

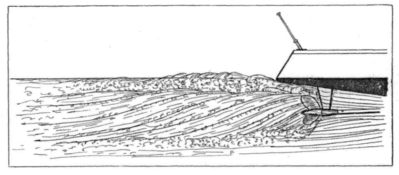

Fig. 55. Action of Submerged Screw Propeller.

or discontinuity in the stream line system necessitating a loss of energy, due to the formation of local eddies.

To make the consideration of the conditions in the immediate neighbourhood of the screw more clear, consider the water stationary and the screw revolving. Because of the suction effect, due to reaction caused by the thrust, the water ahead of the screw is set in motion and flows towards the propeller in stream lines, and constantly accelerates as it passes through the propeller disc, or across the blades, ultimately being delivered sternwards, more or less in the form of a column or jet of water. The propeller, therefore, forms the centre of a stream line system, and when the vessel is in motion the whole system moves forwards with the vessel.

But, as already noted, thrust is impossible of attainment unless there is a discontinuity in the stream line system. This probably commences at the tips of the blades, and in the wake beginning from the screw disc, sternwards.

The acceleration of the water ahead of the screw does *not* contribute to the thrust, but is formed by a circulation of energy from the wake.

The thrust of the propeller is due to the acceleration of the water as it passes through the screw disc, the accelerated portion being discontinuous with the surrounding water.

There is also a rotational momentum equal to the torque, which is probably confined to the immediate wake behind the screw, and the propeller has constantly to impart rotational velocity to new water as it progresses forwards, and this causes one serious energy loss in the screw propeller. In some high speed models this loss is very pronounced, a positive fountain of water appearing behind the screw as the boat passes along. The utmost care must therefore be taken to eliminate this loss, as far as is possible, by reducing the thickness of the blades, fining the edges, and shaping the blades correctly, although nothing can possibly entirely eliminate this loss.

Unfortunately, the actual thrust to be obtained from a screw cannot be calculated accurately from purely theoretical or technical considerations, as so far no formulæ for the calculation of pressures on curved surfaces are available. The average value found by experiment, however, tends to show that only from .6 to .8 of the actual power delivered by the engine to the propeller is given out in the form of thrust. It must be apparent that the problem of the screw propeller is extremely involved, and the results of careful experiments should produce wonderfully increased efficiencies in a model power boat.

It is, however, possible to gain some idea of the approximate thrust to be obtained from every propeller at a given speed from the following formula propounded by Prof. Rankine :—

$$T = A \times S (S-s) \div 5.5,$$

where T = Thrust,

A = Area of propeller disc in sq. ft.,

S = Speed of propeller in miles per hour,

s = Speed of boat in miles per hour,

5.5 = A constant for fresh water.

This is based on the assumption described by A. E. Seaton, M.I.N.A., that the forward motion of the model is due to the reaction produced by the projection backwards of a heavy body with mass equal to the ship at speed of the ship.   Theoretically the quantity of water operated on by any propeller is measured by multiplying the area of the tail race (disc area of propeller) by the mean velocity of flow say in feet per second.

Now assume that A = Area of tail race in square feet.

V is the mean speed of flow in feet per second, 64 lbs. is the weight of a cubic foot of water, and gravity at 32.

   1. Volume of water per second = A × V

   2. Weight   ,,   ,,   ,, = A × V × 64

   3. Mass   ,,   ,,   ,, $= \dfrac{A \times V \times 64}{32}$

$$= 2 \,(A \times V).$$

The acceleration or velocity imparted to it is V—v per second (v = speed of ship).

   4. Then momentum = 2 A × V (V—v) lbs. which is the measure of the propelling force, the reaction being called the thrust.

A *Practical rule* for approximately calculating thrust has been put forward by A. E. Seaton, M.I.N.A., as follows :—

$$\text{Thrust in pounds} = \frac{D \times \sqrt{A} \times V^2}{Pr} \times G$$

A = Area of blades, in square feet,

D = Diameter in feet,

V = Speed of screw feet per second,

Pr = Pitch ratio,

G = Ratio of the distance of centre of gravity of blade face from the boss, to half diameter of blade,

G = in practice, for oval blades, 0.40 ; for broad-tipped blades, 0.46.

A suitable engine speed for the propeller proper can only be designed when the mean pitch has been determined and engine speed settled, and of course there is a wide range of blade shapes, pitch ratios, and projected area ratios to be considered and altered within the limits laid down.

The method of delineating a screw propeller is described in Chapter V, on "Boat Designing."

Much has been said and written with regard to the alleged

wonderful increases in hull efficiency, or speed, by the position of the propeller, its ratio of pitch or diameter, and so forth, but the recently published results of the experimental work of Dr. Froude and later of Mr. W. Luke, at the Institute of Naval Architects, conclusively show the lines upon which propeller efficiency may be increased.

A summary of the deductions set forth by Dr. R. E. Froude on the relation of propellers to hull efficiency show that—

(*a*) A variation of the elements of hull efficiency with variation in speed is generally slight.

(*b*) Variation of number of blades, or pitch ratio of propellers, does not practically affect the hull efficiency elements.

(*c*) Variation of propeller diameter, or its position in reference to hull, does affect these values, and in a manner which we have thus far not been able to reduce to rule.

Mr. W. Luke, at the Institute of Naval Architects in 1910, stated that his experiences practically agree with the above, and were based on two thousand odd experiments made at Denny's Testing Tank at Clydebank.   His conclusions briefly were—

(*a*) Variation in pitch has little or no effect on hull efficiency.

(*b*) Propellers with varying hull clearances show marked gains or losses on hull efficiency.

(*c*) Twin screws when outward turning show a beneficial result.

(*d*) Disc area ratio variation makes little or no alteration to hull efficiency.

(*e*) With single screws, diminishing diameter increases *wakes*, associated with decreasing thrust deduction, a substantial variation in hull efficiency being evident.

Fig. 56. Propeller Action and the Relation of Pitch and Slip.

To grasp the relationship of " pitch," " slip," etc., reference should be made to Fig. 56, which represents a propeller revolving at a constant speed, under varying conditions, and the results to be expected.

AC = Propeller running open or free, no slip or loss.   Conditions are different when propeller is revolving on a travelling carriage, or on a boat, it then only running from A–B.

BC = slip; and $\dfrac{BC}{AC}$ slip as ordinarily expressed as a fraction of the propeller speed, or a percentage by being increased a hundredfold.

But the model boat on its forward way is surrounded or followed by a belt of water with complex motion, and if (over the propeller disc) it can be assumed that its average forward velocity is DB, the propeller is carried along at the forward speed, over the ground, and being revolved as before, would behave in some respects as would an open propeller advancing with speed AD and with slip DC.

Wake velocity is DB.

Wake friction is $\dfrac{DB}{AB}$.

Apparent slip BC, apparent slip ratio $\dfrac{BC}{AC}$.

Real slip is DC, and real slip ratio $\dfrac{DC}{AC}$.

Thrust is appropriate to speed of advance AD, and slip DC.

Thrust, however, is being overcome at a velocity AB, and the apparent work done is Thrust × AB, instead of as with an open screw, Thrust × AD, the excess over unity being the wake gain.

If this were all it is apparent that much advantage is gained by placing the screw in a maximum wake current, and single screw would give maximum advantage ; but there is another point to be considered, viz., Thrust Deduction.

Thrust Deduction is an increase in total hull resistance, because, being caused by a diminution of pressure ahead of the propeller, it virtually increases the resistance, consequently the propeller thrust would have to be *greater* than the total " tow rope " resistance, by an amount sufficient to counterbalance this pressure diminution.

It may be written as an equation thus :—

Thrust $= (I — t) = R.$

That is, T = total thrust, less thrust deduction t, is equal to the tow rope resistance of ship.   Therefore it follows that :—

RV = TV.   (I × W) (I – t).

(Resistance and speed).

The product (I × W) (I – t) is what has been termed hull efficiency.

(W = Wake friction).

T = Total thrust.

V = Speed of propeller through water contiguous to it.

RV   Measures effective horse-power.

TV   Measures effective thrust horse-power.

It follows, then, that—

E.H.P. = T.H.P. × hull efficiency.

The scope of this work and the limits of space prevent a lengthy dissertation on propeller design.   To do such a subject justice it would require a volume of the size of this present book, and consequently those who desire to delve deeper into the mysteries of propeller design are referred to Seaton's " The Screw Propeller " and Prof. Biles' " Resistance and Propulsion of Ships."

Regarding the mere construction of a screw propeller, there are practically only two courses open.   (*a*) To make as accurately as

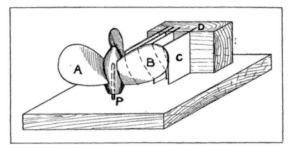

Fig. 57.   A Method of Ensuring Correctness of
Propeller Blade Shape.

possible a wooden pattern, and obtain a casting in brass or other suitable metal, and carefully file and grind it to exact shape.   (*b*) The other method is to turn up a metal boss, slot it at the appropriate root angle, and braze in the blades separately, afterwards hammering or bending them to correct shape, and finishing by grinding and polishing.   In this case the blades could be of steel plate, which would be stronger and lighter than brass or gunmetal, also the balance would be better.

Whichever course is adopted every care should be taken to ensure the actual propeller being *accurate* as regards its shape and balance ; that is, all blades should be exactly alike in shape and weight, so that there is no unbalanced force when the boat is running to cause vibration.

The means of ensuring accuracy are various. One is to cut a hard wood block exactly to shape of desired blade and use this as a guide. Another is to cut several pieces of card or tin plate with their ends truly fashioned, in accordance with the actual drawing of the propeller, and mount these accurately on a base plate, D. A centre pin, P, is then fixed on which the propeller is mounted, the accuracy of blade shape being tested by inspection, noting that each blade, B, lies snugly against each guide plate, C, somewhat as sketched in Fig. 57. The beginner to model power boat work is, however, counselled to buy his propeller ready made, as they can be obtained both accurate and cheap, being made and ground true with special jigs and tools.

CHAPTER V.

HOW TO DELINEATE A MODEL BOAT.

THIS chapter is intended to give the student of naval model architecture a slight insight into a method of designing a model boat, or for that matter a large vessel, as the system is practically the same ; but while the methods to be described may not appeal to everyone, they are claimed as practical, and well calculated to produce satisfactory results. Each separate process in the preparation of a design is described and illustrated, while an example is given for each necessary calculation, and the work has been as far as possible simplified to reduce the labour and trouble in preparing the drawings.   The first consideration is the *purpose* which the model boat is intended to fulfil, and secondly any restrictions imposed by the rating rule, considerations of portability, convenience, or realism of appearance, all of which will affect materially the form of the model, but the method of preparing the drawings will be practically unchanged. There are considerable differences of opinion as to the most suitable form of hull for a given result, and when attempting to design a boat the conflicting factors are found to be numerous. In the first place we are confronted with the desire, or necessity, to reduce the wetted surface, and for this reason we want to design our boat with a semicircular mid section, as this, of course, has the smallest perimeter and the least wetted surface ; but this form gives an exceedingly unstable hull, and on this account we should strive to obtain as flat a floor as possible, this latter form tending greatly to increase the initial and dynamic stability. Such a section, however, would give us hard lines at the bilges, and would not generally be conducive to the best speeds, as it would possess many other disadvantages besides that of the greatest wetted surface ; while to reduce the likelihood of eddy-making we shall need to reduce the draft as much as possible, this factor again increasing the difficulties of designing. Giving due regard to all these considerations, it appears advisable to adopt a midship section with a reasonably flat floor, rising somewhat towards the bilge. The bow, in the case of a racing boat of the " displacement " type, should have easy V sections, the entrance or angles of entry being made as long and fine as possible. For a boat intended for ordinary cruising a somewhat fuller section with increased flare may be adopted, while for purely racing boats the forward section may well be shovel shaped to assist the boat to rise and skate on the surface of the water.

So much for the cross section, but there remains yet the profile or longitudinal sections, and these are generally governed by the size of the propeller, type of engine and machinery. The run aft, however, will be found to test the designer's skill, for a fast boat of the " displacement " type, the water lines at the stern must be sufficiently broad to give the boat ample strength to resist the phenomenon known as " squatting," but to secure the fastest boat and to, as far as possible, reduce eddy-making and wave-making, it is necessary to have the run aft as fine as possible, and consequently it becomes necessary to compromise, and the design of the torpedo boat destroyer represents a practical solution of some of these difficulties. There is, in addition, the question of the power required for propulsion, but this is dealt with more fully in the chapters devoted respectively to machinery and propulsion.

We may now consider the designing of our model, and the processes to be observed are exactly the same for any type of vessel ; although the *purpose* for which the model is intended governs the *form*, the *methods* to be adopted remain substantially the same, and the application of a little care and common sense will enable a satisfactory design to be attempted, although, of course, lengthy experience proves the greatest assistant to a model boat designer.

For the purpose of illustration, and also in view of the popularity of the type and size, the whole process of designing a full sized metre model T.B.D. has been selected, but the lines are far too fine and displacement too small for a practical working model ; such a hull is only suitable for an exhibition model, the lines being given in Fig. 58, and to further guide and assist the designer the lines of several other typical boats are reproduced to a smaller scale. To commence operations :—Secure a drawing board

or other suitable surface and a supply of drawing paper sufficiently
large to enable the construction of the full sized drawing to be under-
taken ; a good straight edge about 4 ft. long, and a few splining
battens with weights for same ; a set of ship's curves, a pair of compasses, set square, tee square, pencils, indiarubber, etc.   If boat
designing is to be taken up seriously, the immense advantages
obtained by the use of a planimeter are sufficiently great to warrant
the expenditure necessary on such a piece of apparatus.   Its use
is to calculate the areas of the cross sections and longitudinal sections
of the vessel, etc., the readings being obtained by the simple process
of reading off the results from the dial, but such an instrument costs
in the cheapest pattern about 45s.   The worker who only wishes
to design one or two boats for his own use, will probably be content
to adopt what is known as " Simpson's Rule," a method of calculation
which gives very accurate results, although somewhat laborious in
use, or " squared paper " may be employed with a fair amount of
accuracy.

Numerous practical articles on draughtsmanship have appeared
in the *Model Engineer*, and to these the student is referred for
further hints on the mere mechanical draughtsmanship.   To
commence our drawing, set up the paper on the drawing board, and
at a convenient distance from the top, in this case about 12 in.
Draw a straight line the whole length of the paper, and mark this
L.W.L. (load water line) Fig. 59.   Draw another line parallel
with this one and about 6 or 7 in. below it for the centre line (C.L.)
of the deck plan of the vessel.   As we have decided the length of the
model is to be one metre, that is 39 in., mark off this distance on the
centre line, divide this into ten equal spaces and draw others at right
angles to the centre lines and number these 1, 2, 3, etc.   Also draw
one water line W.1 $\frac{1}{2}$ in. below the L.W.L. and another water line
W.2 $\frac{3}{8}$ in. below W.1.   These are to show the shape of the water
planes.   Fig. 59 shows the drawing thus far.

When the length of the boat and the volume of displacement
are known, we can by the aid of the prismatic coefficient
ascertain the suitable area of the mid section by reference to Table
No. 1, Chapter IV.   We find that the most suitable prismatic coefficient for such a boat as that now being considered will be about
.55.   We can deduce with the aid of the formula $M = \dfrac{D \div L}{F}$.

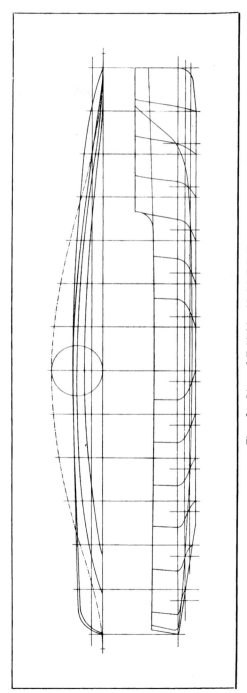

Fig. 58. Lines of Exhibition Model T.B.D.

Where M = Area of midship section in square inches.
              D = Volume of displacement in cubic inches.
              L = Length of L.W.L. in inches.
              F = the prismatic coefficient.

Thus   D = $2\frac{3}{4}$ lbs. $= \dfrac{11}{4} \times 27 = 74.25$ cubic inches.

The displacement of the finished boat being $2\frac{3}{4}$ lbs. it is necessary first of all to ascertain the volume of displacement, otherwise the equivalent number of cubic inches in 1 lb. weight of water. This has been found by experiment to be 27 (nearly), as 1 cub. in. of fresh water weighs .036 of 1 lb.

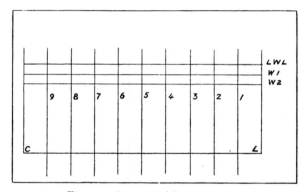

Fig. 59. Preparation for the Design.

Thus $= 2\frac{3}{4}$ lbs. $\times 27 = 74.25$ dividing this by the

$$\text{L.W.L. (39 in.)} \quad \frac{74}{39} = 1.89$$

dividing this by .55, the selected prismatic coefficient, we obtain 3.43 cub. in. as the area of our midship section. Similarly, if (M.) the area of M.S. is known and the length (L.) and prismatic coefficient (F.) the volume of displacement can be ascertained by the formula $D. = M. \times L. \times F.$ Now, on the load water line, and at No. 5, having numbered the cross lines 0, 1, 2, 3, etc., commencing from the bow, which we will assume to be at the right hand side of the paper—extend the line No. 5 for a length of inches equal to the superficial dimension of the area. That is to say, if the mid section immersed area is 3.4 sq. in., the extended line

at this section must measure 3.4 in. from the L.W.L. to the upper
end.   (For large models it is usual to make this part of the drawing
to scale for convenience.) Bisect this line into two parts, and strike
a circle with its base touching the load water line ; this is called the
Generating Circle, G. Fig. 60.   Add to the length of the load water
line forwards about 1½ in. and at the stern 1 in., and then on the forward half of our circle divide the right hand semicircle into four
equal parts from the mid section No. 5, and erect perpendiculars
at these points.   Draw horizontal lines through the four divisions
on the semicircle, and the points of intersection of these lines are on
a curve of versed sines.   Plotting a curve through these points

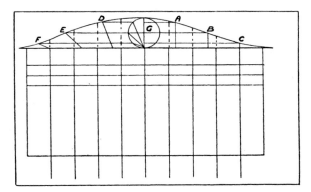

Fig. 60.   The Displacement Curve.

with the aid of the splining batten, our drawing should look
something like Fig. 60 at A.B.C.
    To draw a curve of trochoid for the after body is somewhat more
difficult and requires more care, but proceed in the same manner by
dividing the after half of the generating circle into four equal parts,
and draw chords from these points to the intersection of the midship
. vertical line with the load water line, as shown at D.E.F. in Fig. 60.
Transfer the lengths of these curves, and at the same inclinations
to the four equi-distant points which should have been already
marked off on the after part of our L.W.L. drawing as shown.
Draw a fair curve through the points ; this is called a curve of
trochoid.   The whole curve thus formed as a curve of versed
sines forwards and a curve of trochoid aft is known as a wave form

curve or curve of sectional areas, sometimes called a displacement curve because it represents the distribution of the mass of the hull in the water, or, in other words, it shows the disturbance or displacement of the water caused by the immersed portion of the boat.

Many power boats are purposely designed which do not follow the "wave form" theory, that is, boats whose curve of sectional areas are not based on versed sines and trochoidal curves, but for moderate speed hulls or for preliminary attempts the wave form theory should be followed. The reason for giving an added length to the load water line for the purpose of drawing these curves is because it has been found by experiment that the underwater body can with advantage be increased in efficiency by cutting off the extreme ends of our two curves of sectional areas, this process being known as stubbing. 5 per cent is the minimum amount usually deducted, 15 per cent representing the greatest. If we now measure any lineal distance between the line of our curve of areas and the L.W.L., we shall find that the linear lengths in inches are exactly equal to the immersed area in square inches at that section; it, therefore, remains to design a fair underwater body which shall have its various areas substantially in accordance with these dimensions, to obtain a fair boat with a minimum of resistance at normal speeds, as it has been proved by eminent scientists that the form of the curve of sectional areas in a displacement type of boat is the chief factor governing resistance. We have already decided that the most suitable prismatic coefficient is .55, and from this was deduced the approximate midship area. The next proceeding is to decide upon the form and dimensions of the midship section itself, and for this purpose we construct a full sized trial body plan of the midship section. We know we require 3.4 sq. in. area, and as we are dealing with a T.B.D. we may reasonably decide that our midship section shall have a coefficient of about .7. We would decide upon a beam of 4 in. as a start, this dimension being in accordance with scale size. If we make it simply a rectangle and $1\frac{1}{4}$ in. deep, we have 5 sq. in. of area. Dividing this by .7, we obtain 3.5 sq. in. as the midship section area; this conforms very closely to what we require.

It now remains to draw a form whose area will be equal to this amount. Supposing we try that indicated in Fig. 61. We have already seen that we require for high speed, fineness of lines, and that to reduce the wetted surface we do not want too hard a bilge, but at the same time our little vessel must be quite stable, so we therefore compromise and try a midship section with a reasonably rounded bilge and moderate rise of floor.   With the aid of a planimeter, if available, we can readily find that the area of this midship section is, say, 4 sq. in.   This will be too much.   We now have the choice of two courses, either to simply cut a piece off the top, as it were, removing a rectangular area, the amount of which is easily calculated beforehand, and so arrive at the required result (this is a very practical method, so long as the draft is kept of reasonable proportions), or, on the other hand, we must draw an entirely

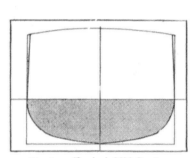

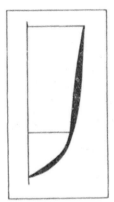

Fig 61.   Typical Mid Section.   Fig. 62.   Trial Bow Section.

new section.   The form we have already drawn is a fair compromise and likely to be satisfactory, so we decide upon the former plan and find by simple arithmetical calculation that it is wise to remove .5 sq. in. from the total amount.   Dividing this amount by four, the width of the beam, we find that the depth of the part to be removed will have to be .125 in.   Strike a line at this point, and again calculate the area; this time the result is 3.5 in., which is sufficiently near for all practical purposes.

Now on the deck plan, set off from the centre line, on section 5, half the width of the beam, and on the sheer plan measure from the load water line upwards the amount of freeboard.   At section 5 this gives us two points upon which to draw trial lines.   Next select

a section about No. 3 or No. 4, and by reference to the curve of areas
we find we require on No. 3 an area of 2.125 in.   The depth forwards
will be about the same as that amidships.   Now draw a trial forward section, the limits of which must obviously be contained in the
rectangle already described, as shown in Fig. 62.   We find that this
works out to 1.9 sq. in.   This, however, is not sufficient according to
our curve of area, we therefore fill it out somewhat as shown shaded

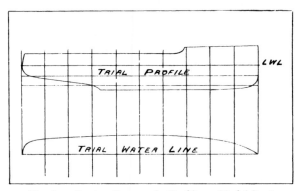

Fig. 63.   Further Progress in Drafting a Model Boat.

in Fig. 62; on calculating again, we find this area to be 2.2 sq. in.,
which is sufficiently near for all practical purposes.   Now at
section 8 we will draw another cross section, but, in this case,
the beam will, to a great extent, be governed by the greatest draft
permissible, so on the sheer plan we will first draw a trial profile,

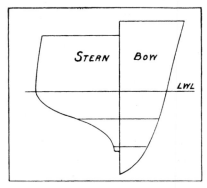

Fig. 64.   Trial Bow and Stern Sections.

somewhat as illustrated in
Fig. 63 ; the point of intersection, of course, gives us the
greatest draft we can adopt.
Dividing this into the area
required, we find that the
maximum beam will not
need to exceed $3\frac{1}{2}$ in., nor
will it have to be much less
than this, as the floor will be
fairly flat.   The section as
illustrated in Fig. 64 may well
represent our first attempt.

which upon calculation is found to have 2.26 sq. in. of area, which is a very satisfactory result.   The remaining sections are comparatively easily filled in by proceeding on similar lines.

A " diagonal " line, Fig. 65, should be drawn, and on it the widths of each section should be marked as shown from the centre line of the body plan, along the diagonal to the outer skin of the boat as shown at D, and these widths set off from the centre line of the deck plan, or from a separate base line, and, of course, this curve must also be fair, that is, it must not have hollows or " humps " in it, but be quite a gradual and easy curve.   The diagonal lines are quite invaluable for rapid " fairing " of the lines of a boat drawing.   To make the relationship of the " lines " of model

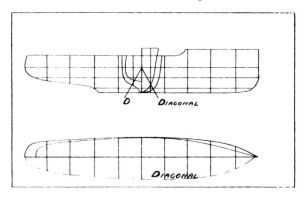

Fig. 65.   Showing the Relation of a Diagonal Section.

more clear, it should be remembered that the water lines are horizontal layers, as if they were sawn off the boat ; the buttock lines are vertical layers ; and the diagonals, as their name implies, diagonals cut through the hull from the centre line outwards.

It is at this stage of the work that the aid of the planimeter will be found so valuable, as without it we are compelled to have resource to Simpson's Rule.   This is a method of calculating an area whose curves are of parabolic nature, and requires that the area to be measured be divided into an even number of spaces, giving an odd number of ordinates.   The area or volume is obtained by multiplying one-third the common interval by the sum of the first and last ordinates.   This is known as Simpson's First Rule.   It will be found advisable to draw out a little table with the sections numbered

and fill in our actual dimensions as we ascertain them.   This rule
is also of value for finding the displacement from our various sectional
areas, and it will serve as a useful comparision, when the drawing
is more complete, to ascertain how near the actual displacement will
be to that indicated by the use of our approximate coefficients,
but we can reserve these exercises for a later period.   Sectional
paper, ruled in one-tenth of an inch squares, provides a ready means
of *practically* calculating the area of the cross sections, and saves
much labour, but the planimeter gives the best and most accurate
results.   To proceed with our drawing : On the body plan draw
a " trial section " for each section line, Nos. 1, 2, 3, etc. Then on the
upper side of the centre line of the deck plan mark off at the respec-

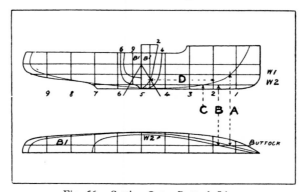

Fig. 66.   Setting Out a Buttock Line.

tive sections the widths of the load water line, as given on the body
plan, and draw in a fair curve through these points, as indicated
in Fig. 66.   That is the plan view of the L.W.L.   Then draw the
water line W.1, which is parallel to the load water line, as shown in
Fig. 66 W.1. To do this mark off the widths of this line (W.1) from
the body plan, and set off these widths at their respective sections
on the deck plan, and draw another water line as fair and true as
possible through the points so obtained.   Now on the deck plan,
draw what is known as a " Buttock Line," ¾ in. from the centre line
and marked B.1 in Fig. 66.   Then on the sheer plan mark the points
of intersection of this buttock line with the horizontal water lines,
and plot a fair curve as far as possible through these points.   With
these preliminary lines to guide us, we can readily fill in the remain-

ing sections, checking off from areas, when it will probably be found that very little alteration will be required to keep the areas in conformity with our wave form theory. A little give and take on the various cross sections and water lines will allow of all the curves being made fair and true, these processes being generally known as "fairing up." In Fig. 66 the buttock is shown wrongly drawn, as the intersection of B.1 on W.2 should be as marked B, whereas the buttock on the sheer plan cuts at C. The correct way in which these lines should agree is shown at A. We should now decide upon the amount of free-board (or the height of the vessel above the water) to give us sufficient height to instal the motor and accumulator or other machinery which

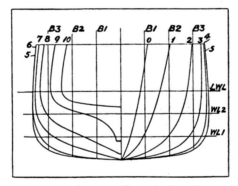

Fig. 67.   Body Plan for Metre Torpedo Boat Destroyer Hull.

measures, say, three inches high, and obviously therefore the model would not be of less total height than this. On the midship section No. 5, mark off this height from the keel line and draw a line curving up somewhat at the forward end, and a slightly less amount on the stern known as the "sheer" or deck line. The trial sheer line already drawn will probably be found to be substantially correct. This will give us a nice sheer line or deck line as shown in Fig. 58. On the deck plan we then set off the widths of the sections at the points of intersection already found, and complete the body plan shown in Fig. 67 ; the sections to the right being those forwards, and those on the left being the after part of the vessel. This body plan is simply drawn for convenience, to enable us to readily ascertain the general form of the hull, of which the heights and breadths will have to be carefully checked to insure their being in agreement at the appropriate spots, the result, if carefully drawn, being very similar to Fig. 67, which shows complete lines for our model torpedo boat destroyer.   Other chapters give a further detailed description of the fittings, but we have yet to decide upon the position of the motor, accumulator, or steam machinery and also the position of the propeller and shaft, to insure our model floating on an even keel, and in addition to ascertain that our boat will be stable, and further that the actual design is of correct displacement.

A study of Fig. 58 will reveal two features that are apparently erroneous.   Firstly, twelve sections are shown in place of the nine referred to in the foregoing remarks, but this is quite immaterial, as the example of drawing here described is to show the principles on which any power boat may be designed, while to give greater accuracy in the building of the hull a greater number of cross sections are shown in the plate, but the area curve once drawn as described herein, and as shown by Fig. 60, it is perfectly simple to ink in the "curve of areas," and then rub out all the pencil constructional lines and draw the twelve (or more) cross sections and measure each one for " areas " as described above, but the nine sections are very useful as by their aid it is much more simple to calculate the C. of B., etc., as will be shown later.   The second feature is the impracticability of the design as a *working* model.   The lines in Fig. 58 have purposely been drawn to correct scale to demonstrate that it is not necessarily sufficient to merely reproduce in miniature the lines of an actual ship and expect them *necessarily* to be good for a working model, but on the contrary the practical working model boat must be designed throughout as such.   Thus to get a reasonably large working displacement the beam and draft must be increased, and the fullness of the boat generally increased in a like manner, and the torpedo boat destroyer shown in Fig. 68, which gives the lines of a similar class of boat to that described, but with a greatly increased displacement, serves as an excellent example and clearly shows the lines on which practical small model power boats differ from their so-called prototypes.   For further assistance to the amateur designer other examples are given in this chapter of typical model boat hulls, notably Fig. 69, lines for a model battle cruiser, and Fig. 70, the lines of a practical tug boat model.   These designs have all been built to and proved very satisfactory in practice.

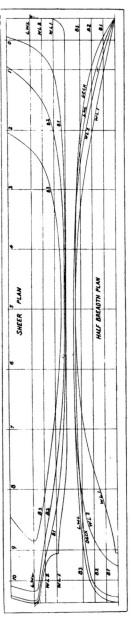

Fig. 68.   Lines for a Metre T.B.D. Hull.

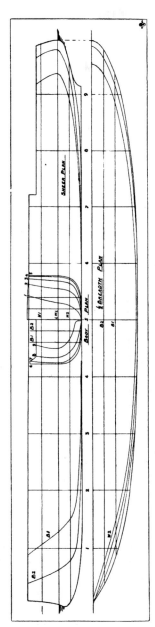

Fig. 69.   Lines of a Model Battle Cruiser.

The further consideration of the design of any model power boat will now be continued, and to prove the design and ascertain from the drawing the actual displacement of the model, proceed to calculate from the body plan the areas of each immersed section, either by Simpson's Rule or by measuring direct from the area curve, if correctly drawn, and then—where there are an *odd number* of sections evenly spaced—proceed as follows by putting down in tabular form the areas of each section and multiply by Simpson's multipliers as follows :—

TABLE No. 4.

CALCULATION OF DISPLACEMENT BY SIMPSON'S RULE.

No. of Section.	Area of Immersed Section.	Simpson's Multipliers.	Product for Volume.
1	·5	1	·5
2	1.125	4	4.5
3	2.125	2	4.25
4	3.0	4	12.0
5	3.1	2	6.2
6	3.33	4	13.2
7	2.8	2	5.6
8	2.2	4	8.8
9	·75	1	·75
			Total 55.80

The common interval being 3.9 in., one-third common interval is 1.3 in.   The total volume of displacement is found by multiplying 55.80 by 1.3 = 72.54 cub. in. = 2.68 lbs.

As will be seen above, the " areas of sections " must be multiplied by Simpson's multipliers, thus giving the product for volume. Multiplying the total of these " products for volume " by one-third the common interval, gives the actual area of the immersed portion of the hull, and dividing by 27 (the number of cubic inches of water in 1 lb. weight), gives the weight in lbs. of the " volume of displacement," otherwise the total weight of the boat, with all machinery, etc., on board ready to run.

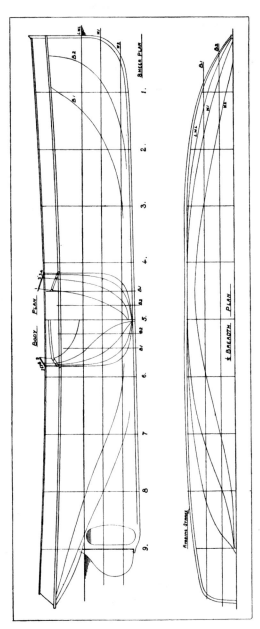

Fig. 70. Lines of a Model Tug Boat

H

Providing the design has been produced with the areas of each
section in conformity with the " curve of sectional areas," drawn
at the commencement, it is obvious that the displacement can be
calculated *before* the lines are drawn, and also the position of centre
of buoyancy ascertained.   This is a great boon, which is appreciated
when considering the merits of several types of hull, etc., as the
areas can be measured along each section line from the centre line
to the " curve of areas," this *linear* length giving us the superficial
area, or area in square inches of the immersed section at that point,
and can thus be filled in to the Table 4 direct.   Of course if the
" curve of areas " is drawn to scale the length on the drawing must
be multiplied accordingly.

The fore and aft position of the centre of buoyancy is found by
using the areas of immersed sections as in Table 4, and taking
moments about the midship section.   In this case the products
of the sections before are subtracted from the moments abaft
the mid section.   The " products for volume " are multiplied by their
distance fore or aft of the mid section.   Thus :—

TABLE No. 5.

Fore and Aft Position of Centre of Buoyancy.

No. of Section.	Product for Volume.	No. of Intervals between Section and M.S.	Product for Moments.	
1	·5	5	− 2.5	
2	4.5	4	− 18.0	
3	4.25	3	− 12.75	
4	12.0	2	− 24.0	
5	6.2	1	− 6.2	
6 (M.S.)	13.2	0	—	
				− 63.45
7	5.6	1	5.6	
8	8.8	2	17.6	
9	·75	3	2.25	
			25.45	
				− 38.

Common interval between sections is 3.9 in.

C. of B. is $\dfrac{3.9 \times -38}{55.80}$ — 2.65 inches from the stem of the model.

The vertical position of the C.B. can be calculated in a similar manner, but taking the waterplane areas, as shown by the water lines. For practical purposes, however, the C. of B. may be taken as being .4 of the mean draft below the L.W.L.

The reserve of buoyancy is the buoyancy of all water-tight portions of the model above the load water line, and depends upon the form of the above-water hull, and upon the height of the freeboard. Having ascertained the fore and aft position of the centre of buoyancy, the *trim* of the model can be calculated, and the correct placing of the principal weights ascertained. Of course, when the hull is already made, there is no advantage in calculating

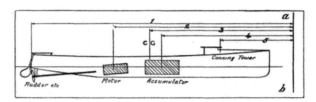

Fig. 71. Finding Position of Centre of Gravity.

the " trim," as a few suitable lead weights to represent the machinery, or the actual machinery itself, may be placed in the hull while it is floating in water, and moved forwards or backwards until the desired trim is found. But in designing a new boat it is imperative to know with considerable accuracy the places for the more important weights, or pieces of the machinery. The correct position for the sum or total of all the weights, including hull, rudder, engines, boiler, deck fittings, etc., is such that the mean centre of gravity of all these weights is perpendicular with the centre of buoyancy. To ascertain this it is necessary to " take moments " from a neutral line adjacent to the hull, as shown in Fig. 71.

Draw a line *a b* at a reasonable distance from the bows of the boat. Take moments from this line by multiplying the lineal distance in inches as at 1, 2, 4, 5, by the individual weight of each part of

the boat in ounces.   For practical purposes the hull weight may
be assumed as uniform, and if the chief weights, *e.g.* motor,
accumulator, deck fittings, etc., are calculated, the result should
be quite satisfactory.   The sum of these products divided by the
number of weights gives the " mass distance," if it may be so
termed, that is, the sum of the *weights* multiplied by the *mean*
distance (3) from *a b*.   The total weight being known, divide the
" mass distance " by the total weight, and this gives the position of
centre of gravity.   If this is ahead or astern of the centre of
buoyancy the weights must be adjusted to suit, until the centre
of gravity is perpendicular to centre of buoyancy.

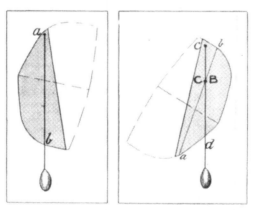

Fig. 72.   Fig. 73
Practical Method of Finding Position of Centre of Buoyancy.

The stability of the boat must now be proved, and for ordinary
model power boat work, if the stability as found by means of the
mid section is satisfactory, the boat may be considered as all correct
But for accurate work the total alteration of buoyancy of the total
immersed sections of the hull should be taken.   To ascertain practically the stability of our model, commence making a full size drawing of the mid section.   Mark on this the vertical height of the centre
of gravity (C G), which may be calculated in a similar manner
to that adopted for the fore and aft position of the centre of
gravity but measured from a line parallel to the load water line
Then mark the position of the centre of buoyancy (C. B).

Now by reference to Chapter III we know that when the boat is inclined the position of centre of gravity remains unaltered, but the centre of buoyancy moves to the side of the model that is immersed. Consequently, to find this new position of the centre of buoyancy when inclined, we should properly draw a complete set of the cross sections, and calculate the mean centre of buoyancy of this altered *form* of underwater body ; but for all ordinary model power boat purposes the mid section will suffice. To ascertain the centre of buoyancy of the altered immersed mid section, we must ascertain its mean centre, which can be done by means of a piece of smooth card, cut to shape and balanced vertically from one corner upon a horizontal pin point with a thread and weight attached —draw a pencil line where the cord is hanging as *a b*, Fig 72, and then place the pin at another corner, and mark the new line

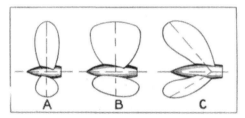

Fig. 74. Three Examples of Propeller Blade Shapes.

c d, Fig. 73—where these two lines cut is the new centre of buoyancy of this section.

If now a full size mid section is drawn and the new centre of buoyancy as found is outside or beyond the centre of gravity the boat is stable ; if not the beam must be increased, or the centre of gravity reduced. This is a very quick, simple, and practical way of testing the stability, and with any normal type of hull is quite sufficient.

Having now completed the lines of our model, the next consideration is the selection of a reasonable propeller, suited to the power available and the resistance to be expected from the hull.

We first calculate the probable resistance as detailed in Chapter IV, and then select the type of propeller most suited to it. Suppose for example the propeller is to be 3 in. diameter and the speed of the boat to be 10 m.p.h. This equals 886 ft. per minute. The engine speed we will assume to be 2500 r.p.m. and slip 15 per cent., the mean pitch of the propeller must therefore be 5 in. As regards the proportion of blade area to disc area, and shape of blades, many designers have a fancy for a particular form of blade, but the elliptical blade as used in the Navy, A Fig. 74 is a good all-round type, while for small high speed engines the broad tipped oval

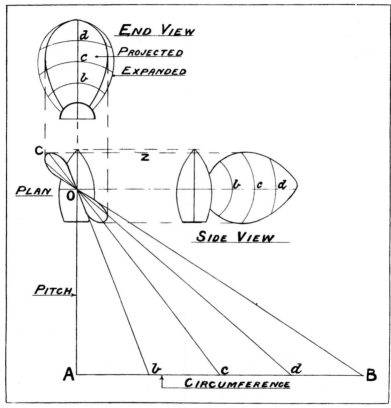

Fig. 75. Drafting a Simple Propeller.

blade B gives good results. For racing boats a raking blade C is very popular; it has the merit of increasing the blade area and length without increasing the diameter, and to a certain extent diminishes the rotational fluid pressure losses. Whatever the type of propeller the method of drafting it may be summarised as follows: Fig. 75

shows the methods of drawing a propeller with vertical blades such as Fig. 74 A and B.

In Fig. 75, A—B is the circumference of the propeller at the tips and A—O is the pitch ; divide A B into four as at *b c d* and join up these points to O and produce the lines beyond as shown.   Then angle A O B is the angle of the blade at one fourth the diameter ; similarly A O *c* is blade angle at half the diameter and so on.   Then

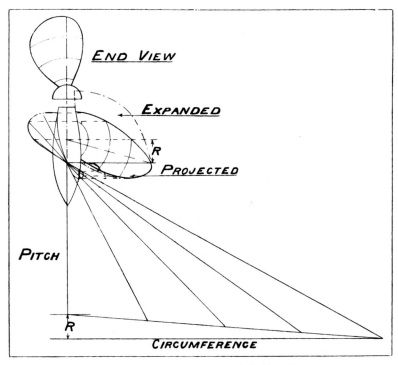

Fig. 76.   Drafting a Propeller with Raking Blades.

at the upper part of drawing draw the proposed developed area of the blade as shown.   On this strike arcs, as shown, dividing the blade into four, similarly to line A B and mark off the width on the projected line C—O, and an equal amount the opposite side of centre line.   Treat all the other lines in similar manner and join up these spots with a fair curve as shown ; this represents a plan view of the blade and boss.   Projecting lines vertically upwards to intersect with the curved lines *b c d* gives the spots for the curve of the " projected area " of the blade, and by projecting a blade horizontally as at Z the vertical or side view of the projected blade is shown.

The method of drafting a raking bladed propeller is given in Fig. 76, which is similar to above, but the angularity of the rake is given to the generating line representing the circumference as marked R.   Except for show purposes the method of drawing either a two or three bladed screw is the same as that given above.

A few points to constantly bear in mind with propeller design is to reduce the blade thickness and provide a reasonable size boss with long tail nut.

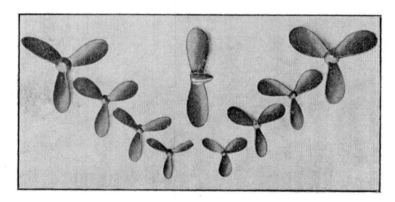

Fig. 77.   Some Typical Model Propellers.

Fig. 77 shows some typical two and three bladed propellers of excellent design and manufactured by Bassett-Lowke.

As regards the relative merits of the two and three bladed propellers, this all depends on the purpose for which they are intended, but generally for very high speed work two blades give best results, while for general all-round use the three bladed propeller is still the best.

For many reasons the equivalent speeds in feet per minute and miles per hour are required ; they are given in the following Table No. 6, and will be found handy for calculations and also in timing a model over a known course when record breaking.

TABLE No. 6.

SPEEDS IN MILES PER HOUR, FEET PER MINUTE, AND FEET PER SECOND.

Speed in Miles per hr.	Feet per min.	Feet per sec.	Speed in Miles per hr.	Feet per min.	Feet per sec.
4.6	405	6.75	14.1	1,241	20.69
5.1	456	7.59	15.2	1,342	22.38
6.0	532	8.86	16.1	1,418	23.64
7.1	633	10.55	17.2	1,520	25.33
8.0	709	11.82	18.1	1,596	26.60
9.2	810	13.51	19.0	1,672	27.87
10.0	886	14.78	20.1	1,773	29.55
11.2	988	16.47	21.0	1,849	30.82
12.0	1064	17.73	22.1	1,950	32.51
13.2	1165	19.42	23.0	2,026	34.20
			24.1	2,128	35.47
			25.0	2,204	36.73

In concluding these notes on the technical aspect of model power boat design, much has necessarily been left unwritten, but it is hoped enough has been put forward to enable the reader to grasp the fundamental fact that model power boats can be and should be scientifically designed—and further that these imperfect explanations will induce the reader to think and consider the why and wherefore of things, and thereby attain to a higher understanding of his pastime and obtain from it the utmost knowledge and satisfaction.

CHAPTER VI.

CONSTRUCTION OF HULLS.

THERE are four practical methods of constructing a model power boat hull, which may be classified in their order of merit.

1. The bread and butter, or superimposed planking.
2. The dug-out, or solid block.
3. Rib and plank.
4. Metal hulls.

The first and second methods are by far the most popular and successful. The dug-out method is employed for boats of moderate size, a solid block or blocks being cut to shape, and requiring practically no internal ribs for strength. The other method, of superimposing planks of wood one on another, and each about one inch or more in thickness, is very often adopted, and provides a light and easily made hull.

The third method of building a boat, with ribs and longitudinal planks, is, as a rule, only used for high-class boats, but if the workmanship is good it makes an exceedingly light and exceptionally strong hull.

The fourth method of making hulls, from metal, has many adherents, but, although it is possible to make a motor boat of reasonable type with the utmost ease, by simply bending a piece of tin or zinc, and hammering the same somewhat near to a true shape, there is little doubt that the metal built boat requires far more skill to construct, and exceedingly good work to produce a model boat hull fair and true; and although the author has seen some remarkably fine pieces of such work, the beginner is strongly counselled to adopt either the first or second method.

The actual methods of building the hulls vary somewhat, but many of the processes are similar, and these will be dealt with one by one. As regards the tools necessary, so much depends upon the knowledge of their use that it is difficult to give a specification that will meet all requirements ; but a handsaw, keyhole saw, spokeshave, small plane, some half inch and one inch chisels, a selection of flat gouges, a brace or hand drill, glue pot, hammer and screwdriver, may be considered among the essentials, also a strong table or work bench. A fretsaw is a very useful tool, and if much cutting is to be done a frame saw should also be obtained. For the sake of simplicity, we will assume we are about to construct a model cross channel steamer. For this we shall require two blocks of the very best yellow pine, and no trouble should be spared to obtain some sound stuff, to get it dry, and free from blemishes or knots of any description, as so much of the easy and successful construction of the model depends upon the materials to be used.

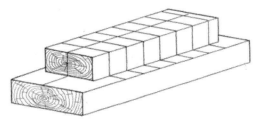

Fig. 78. Marking Out Two Planks.

These pieces when planed up should measure 3 ft. 3 in. long, 6 in. broad, and 2 in. deep, and for convenience of reference we will term the uppermost block A, and the under one B. Strike a centre line down the centre of each piece, and scribe a line every 3.9 in. or equal to the lines on the drawing, and at right angles to the centre line, and number them, as shown in Fig. 78. Take great care over this, as much depends upon this initial marking off. Having pro- ceeded so far, lay this part aside, and prepare the " moulds." These may be made from cardboard, and form templates or guides to the accurate shaping of the hull. Procure some thin cardboard, the kind sold at a penny per sheet by most stationers will do, as it is white and of sufficient thickness. Mark off these sheets of card- board, and cut out ten rectangles, each measuring 5 in. wide by 6 in. high. Great care must be taken to get two edges quite square, as otherwise the hull might, when finished, be found to show a twist or wind.   Mark these two edges A and B to prevent error.

Reference should now be made to the full-sized body plan of the model, and it will facilitate matters if a tracing is made from the drawing, as we have to " prick off " from the body plan on to the sheets of card already cut the outline of the respective sections; this process being simplified by employing a tracing. The method, in any case, is to lay under the drawing one of the pieces of card that have been prepared, taking care to have the edge of the card marked A touching the centre line, and the edge B touching a base line on the drawing.   There is no difficulty with this, as the edge of the card can be easily felt by passing the fingers across these lines, and adjusting the card to its correct position; but this is still easier if a tracing has been made, as the card is then visible beneath it.   Two drawing pins or tacks should now secure the drawing and the card to the work bench to prevent either from moving.   Then prick off the midship section, and for this purpose a sharp scriber or awl may be used; but take care to use a sharp instrument, and one that is not too thick.   The point should be set first on the base line where the curve starts, and the curve will have to be pricked every half inch.   On moving the card it will be noticed that a series of pin holes have been pricked through the drawing.   Cut out the card with a sharp knife through the dots, and carefully remove the centre portion; mark both parts with the appropriate section number, and keep these, as they will be useful later on.   The edge of the card can best be cleaned to shape with No. oo sand-paper, and every care must be taken to see that the curve corresponds exactly with the drawing. Proceed with all the other sections in the same way, and do not be satisfied until the card template or mould fits exactly to the drawing when laid upon it.   The position of the load water line and the sheer or deck line should also be marked on these cards. When all the moulds have been cut to shape and size, we may take the two blocks in hand again, and mark off on the upper block A, at each appropriate cross section, the correct width of the deck, this measurement being readily obtained from the deck plan; these widths, of course, being equal to those on the body plan.   Then with the aid of a splining batten a curve is drawn through these points, this giving us the greatest width on the deck line. Now turn the block upside down, and draw other centre and cross lines to correspond with those on the upper face, and mark off the width on this water line, the widths being given on W. L. in the drawings. Parallel to this line, but at least ⅝ in. inside it, draw a curved line, and after checking all the dimensions over again to prevent errors, cut the block and plane it up true to the shape of the deck plan; afterwards, with the keyhole or frame saw, cut away the inner portion of the block, by this means saving a great amount of labour. The under block B is now marked off, using the widths as on the water line No. 2, afterwards cutting the block to shape and planing it true to this line. We now have two blocks roughly boat-shaped, which require firmly fastening

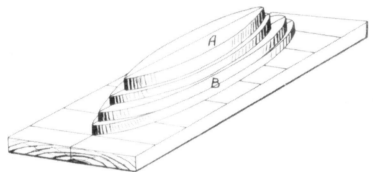

Fig. 79. The Hull Roughly Cut to Shape.

together before the final shaping. There are many methods available for the joining process, but a good glued joint made with one of the patent waterproof glues, and strengthened with a screw or two fore and aft, will prove amply strong enough for ordinary medium sized models; if anything stronger is required, some modification of the bread and butter, or sewn systems described later must be followed. The rough hull must now be laid aside for at least twenty-four hours to dry hard and solid.

To facilitate the external shaping of the hull it is advisable to use a " building board," and for this procure a piece of thick board, sufficiently large to accommodate the hull, and having planed up one side, and marked the centre line and cross lines in exactly the same way as the block A was originally marked, screw the hull, keel upwards, on the board, with the centre and cross lines in agreement, somewhat in the manner shown in Fig. 79, which shows a hull in four layers. The next process is to correctly shape the midship section, so take the midship mould, and apply the edge B of the mould to the midship section line marked on the base board, and cut away the surplus timber along the whole length of the hull, until the mould fits fairly well into its place, that is to say, the edge B must stand square and flat, and the curve of the section must fit to the hull when the edge A of the mould reaches the centre line on the hull. Take care not to cut away too much at the commencement, and do not lose sight of the centre line on the hull. Treat both sides alike, fitting sections 7 and 8 fairly well to their places, and afterwards fitting the forward and

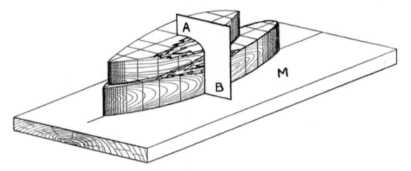

Fig. 80.   Fitting Card Mould to Wood Block.

after sections alternately. Chisels and gouges will be the most useful tools for this work, but the plane is generally used to shape up the midship section. Do not forget to work from the mid section, fore and aft alternately, as this method ensures the wood cutting cleanly and with the grain.

It is hardly possible to give written instructions to absolutely cover all possible contingencies in model yacht building, but with the aid of Fig. 80 some idea of the appearance of the model about this stage can be obtained. This shows some parts nearing completion and a mould fitting nicely. The cross section lines both on the hull and building board are clearly visible, and are numbered to agree with the drawings as indicated. The mould is drawn at M, when the value of the rectangular card is very clearly shown,

the edge B applied to the base board, which equals the base line on
the body plan, and edge A representing the centre line : thus the
only thing to do is to cut away the surplus wood until the mould
stands true with edge B flat on the base board, and edge A in line
with the centre line on the hull.   The deck line and water line 2,
being already planed to size, give us an additional guide to the boat's
shape, and should not require cutting at all.   When all the remaining sections have had their appropriate moulds fitted, and the
intervening timber cut away fair and smooth, the hull may be
unscrewed from the base board, and the hollowing of the interior
proceeded with, gauging the amount to be cut away by using a
large pair of calipers, or by cutting a good half inch off the curved
part of the inner portion obtained from the moulds, using these

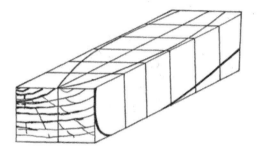

Fig. 81.   Marking Out Solid Wood Block.

for the inside of the hull, in a similar manner to that adopted for
shaping the outside.   Take care not to cut away too much timber
inside the hull, and leave an extra thickness of wood round the
screws, holding the two blocks together.   The " sheer " must now
be " cut in," and this is easily done by marking the depths from
the body plan, measuring from the load water line upwards, and
setting off these heights on the hull, and afterwards joining them
into a fair curve with the batten.   Cut away the hull to this line,
using the spokeshave, or a small plane, and the hull is ready for
the final finishing with sand-paper, and a coat of " priming " colour
completes this part of the work.

    The foregoing method of boat construction applies practically
to either the dug-out or bread and butter systems.   In the former
case a true dug-out hull would be one made from a solid log of

wood. This method is, however, very laborious, and really possesses but few advantages. With the dug-out method, the solid
block is first planed up true to shape, size, and rectangular in form ;
it is then " marked out," Fig. 81, as before described, sawn to the
shape of the deck line, and simply fashioned by hand tools with
the moulds as a guide, the interior being laboriously dug-out
by hand. While doing this it is convenient to clamp the hull to
a table, as shown in Fig. 82. The advantage of the two block
system should therefore be apparent.

The bread and butter, or " laminated," system is a development

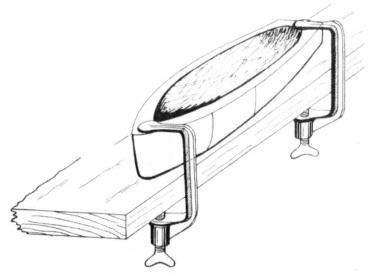

Fig. 82.   Hollowing Out the Hull.

of the two block system, and usually a larger number of thinner
planks are used, thus ensuring greater accuracy and the most rapid
construction. Other advantages of this system are economy, as
the centre portion of the upper planks can usually be cut to shape
for a lower layer.

The method of construction is practically as described previously
for the two piece system, but all the required planks must be planed
upon both sides perfectly true, taking great care to make them of even
thickness throughout. Proceed by marking the centre line down the

middle of each plank and marking off other lines at right angles. Number all these 1, 2, 3, etc. These correspond to the sections of similar number on the body plan. Next mark off each of the water lines, measuring from the centre outwards with a pair of dividers to the spot where the section lines cut the horizontal water line as shown in Fig. 83 at C; set off this distance C on the same cross line on the first plank, and do the same for all the remaining sections of the first water line, then with a curve or batten D, bent around a series of pins or brads driven in on the spots marked on the cross lines, draw a fair curve which represents the shape of the boat at the first

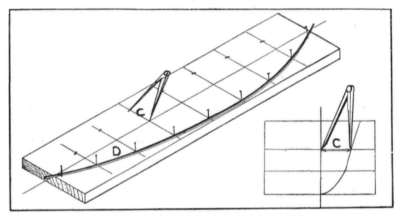

Fig. 83.   Setting Out Plank for Cutting to Shape.

water line, and do the same for each of the other water lines, when the planks may be sawn to shape and planed up true to these lines. Place them one on top of the other upside down, that is, so that the bottom of the boat is uppermost. Take care that all are true with regard to the centre line, and that all the cross lines coincide, and run a pencil around each plank as shown in Fig. 84, then mark off another line ⅜ in. or more inside this line, making it a little stouter at the bow and stern, and saw out the interior. Thoroughly warm the timber, and with some best Scotch glue, thoroughly hot, and having 10 per cent. solution of bichromate of potash dissolved in same, glue these planks together and place heavy weights on top of them or screw them up with carpenter's cramps if same are available.

The greatest care should be taken in gluing these planks together

I

to see that the centre lines and section lines are in their correct places. After the glue is thoroughly set, which will take at least two days, the outside of the hull may be finished with the aid of cardboard moulds in the ordinary manner, and the inside hollowed out as light as possible, the joints being stitched with copper wire as shown in Fig. 85. This stitching is of great help in maintaining

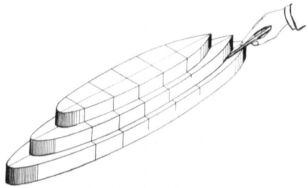

Fig. 84.   Marking Off the Water Lines.

a sound joint, as without it there is considerable risk of the joints " starting," due to the vibration of the engines.   The wire used for the stitching may be of medium thickness copper, about $\frac{1}{16}$ in. dia. will be suitable.   This is bent into a ⊔ shape, and "stabbed through" the hull with one leg of the ⊔ through the upper layer

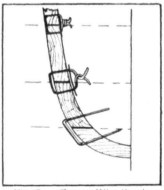

of wood, and the other leg through the lower ; the ends should project into the hull about $\frac{7}{8}$ in. for 3 ft. to 4 ft. 6 in. boats.   The ends are then slipped into a short piece of perforated zinc, and twisted up tightly, driving the copper stitch tightly into place, by tapping the outside portion with a hammer, until the wire has sunk right into the wood of the hull, when the joint will be found to be quite sound and tight. Of course, glue should be used to

Fig. 85.   Copper Wire Stitch.

cement the joints, the stitching being added when the hull is practically finished. The stitches should be spaced about three inches
centre to centre for average work. The outside of the hull should
be stopped with good red and white lead in equal proportions,

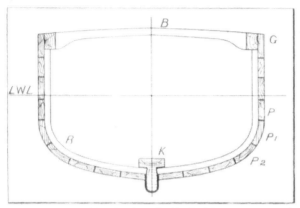

Fig. 86. Section of a Planked Boat.

cleaned off flush when dry. The whole of the hull, inside and out,
should then receive a good coat of lead priming colour, and is ready
for the machinery. Of course, the final painting must be done after
all the other work in the hull has been completed.

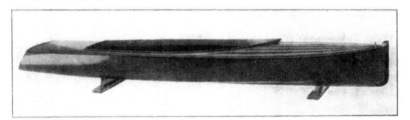

Fig. 87. Mr. C. Coxen's Motor Boat *Zingara*.

The method of model boat construction known as "plank
building" consists in building up the model in similar manner to the
full size boat, by planking the hull with thin cedar or mahogany
planks secured in shape and position by bent or sawn ribs. By this
method the backbone or keel of the boat is first cut to shape, and

the stem and stern posts cut out and fitted in place; the ribs are
then bent and secured to the keel, about three inches or so apart,
and the planks afterwards cut, fitted, and secured in position.
A cross section of a typical plank built boat is shown in
Fig. 86 ; the backbone of the boat is the keel K. Ribs of American
elm bent to shape are shown at R, the planks and their modes of
fastening being shown at P, Pi, P2.   It will be noted that the ribs.

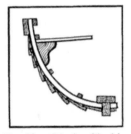

Fig. 88.   Clincher Planking.

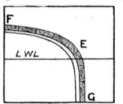

Fig. 90.   Showing Allowance for Planks
when Cutting the Mould.

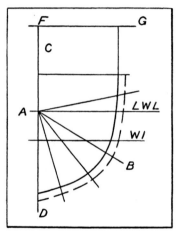

Fig. 89.   Setting Out the Mould
Plan.

R, are secured to the keel, K, and that this is recessed to receive
the inner plank, or "garboard strake." Such a model is usually
lighter than the dug-out, or bread and butter hulls, and
although the system of plank building has only been extensively
used for model sailing yacht construction, the system can with
advantage be applied to model power boats.   The work of Mr.
Charles Coxen, of Southsea, the well-known amateur boat builder,
will serve as an excellent example of what can be done in this
connection. Fig. 78 shows one of his 1½ metre hulls, the *Zingara*.
In the space available, only the general principles of plank building
can be described, as it is impossible to foresee every difficulty with
which the constructor may be confronted; the general principles

of the work being noted chiefly, so that each individual can adapt them to his own particular needs.

For convenience, the description of a model launch hull will be adopted, but the following method of hull construction is equally applicable to any model, either sailing or power, with but very slight modification. The easiest forms of hull for plank building are those with rounded or " easy " sections, and such a design should be selected for this reason. There are several systems of plank building, the chief of which are—

(a) The " Clincher," such as employed for ordinary rowing boats, where the planks are clenched one to the other, by nails driven through the overlapping planks (Fig. 88).

(b) The " Carvel " or smooth skin construction, where the joints are kept water-tight by reason of the well-known property of some woods to swell when placed in water. The use of *dry* timber is therefore imperative for such work.

(c) The diagonal and double skin : this is not used to any extent for model work, the " Carvel " system being almost universal, and it is this system that is described in this chapter.

Having obtained the design, the first process is to construct a " mould plan," a drawing similar to the body plan, but smaller by the thickness of the planking and the ribs. For small models up to 5 ft. in length, the planks may safely be made $\frac{1}{8}$ in. thick, and the ribs, if bent, may be also $\frac{1}{8}$ in. thick, but if they are sawn they must be somewhat thicker, the amount depending on their shape and position. For ordinary work the bent ribs give excellent and satisfactory results.

The mould plan may readily be drawn from the body plan by constructing a diagram on tracing cloth such as is shown in Fig. 89 the centre line, C–D, load water line, L.W.L., and the water line, W1, being drawn exactly as on the body plan. The lengths from the centre A to the section line as at B are then marked off along the corresponding line on the mould plan, less the thickness of the rib and plank, in this case $\frac{1}{4}$ in.; the true section of the outside of the boat when planked is shown at E, Fig. 90, the " mould section " being shown by F G. Each section is treated in a similar way, except the transom, or end plank, which is only $\frac{1}{8}$ in. (the thickness of the planks) smaller than the body plan, and section No. 1, which is also solid, and consequently has only to be made $\frac{1}{8}$ in. smaller than the body plan, as no ribs are fitted at section No 1, a solid bulkhead being used instead.

Having marked off the shape of the moulds draw a parallel line as at F G Fig. 89 to form a base line at right angles to the centre line C D parallel to the L.W.L., and at a convenient distance from the same, say six inches.

The next process is to cut the transom, from ¼ in. mahogany or pine, and the bulkhead No. 1 from similar material, using only the very best and dry timber, as although the best selected stuff is expensive, such a small quantity is required that it proves most economical in the end.   If possible, use old dry wood that has been in stock in a good dry place for a year or two, as in this class of work shrinkage of the timber is most aggravating and objectionable, causing the seams to open and leak.   The moulds themselves may

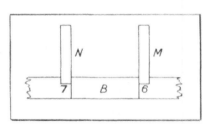

Fig. 91.   The Moulds Mounted on
Base Bar.

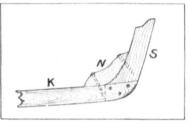

Fig. 92.   The Kee and Stern
Piece.

be cut from any rough material about ⅜ in. thick, but the centre line, C D, load water line, L.W.L., and the base line, F G, Fig. 89, must be marked clearly and accurately on each mould, and the curved sides cut to correspond, so that when all the moulds are cut to shape, according to the mould plan drawing, and stood upright on their base, F G, the L.W.L. lines will agree and make a straight line, as well as the centre line, otherwise the boat cannot be true and fair.   These moulds are now to be mounted on a bar of timber about 3 in. × 2 in. in section, and planed up true, the moulds being screwed in place, and the correct distance apart, as shown in Fig. 91, the forward sections Nos. 2, 3, 4, 5, 6 being screwed so that they are in front of the section lines, and the after sections, Nos, 7, 8, 9, etc., being screwed in place so that the thickness of the mould is behind the section line, B being the base bar, M the forward section mould with their

thickness in front of the section lines, and N the after section mould
with their thickness behind the section lines.   Having secured the
moulds with screws as shown, take a batten about three feet in length
and ¼ in. square, and lay it around the moulds, when it will be noted
that if the batten is bent fairly easily it will lie on the front edges
of the forward sections M, Fig. 91, and on the after edges of the
after sections N.   To complete the shaping of the moulds these
edges must be cut to a bevel to correspond with the shape of the boat
as indicated by the batten, so that when the ribs are in place the
planks will fit fairly and easily to them, and not lie on one edge only
of the rib.   The moulds may then be laid aside while the keel is cut
from ¾ in. thick mahogany.   The shape and dimensions of this keel
piece, of course, vary with various models, but the principle remains

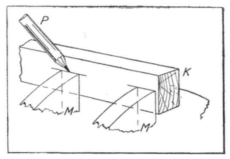

Fig. 93.   Marking Off the Keel.

the same throughout.   When this part has been cut to shape, cut
another piece of the same material for the stem S, Fig. 92, and another
piece for the knee, N.   The stem, S, is then morticed into the keel
K, and the knee N glued and screwed into place, making a good
sound joint.

When the glue has set thoroughly hard plane up the piece perfectly
flat and true and fit the transom by screwing it into the keel.
The bulkhead No. 1 is also screwed in place and both these parts
may ultimately be glued, but it is best to leave this until all the work
on the keel has been completed, and the ribs are ready to be fitted
in their place.   The mould box must now be taken in hand, and an
upright fixed at the end section and arranged to act as a " fence " or
stop for the transom, to ensure it being in its correct position as
shown in the drawings, another block for a like reason being fitted

to support the transom.   If the keel piece is now tried in place, it will be found that the mould must be cut away in the centre to allow the keel to fit, so carefully cut out the centre portions of the moulds until the keel can fit into its proper place, when it will be found to project above the moulds by approximately $\frac{1}{2}$ in.   Now with a pencil mark the positions of the moulds on the keel as shown

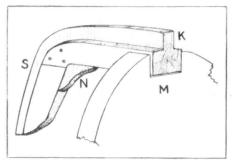

Fig. 94.   Cutting Rabbit in Keel.

at Fig. 93.   These places indicate the bottom surface (as the mould stands) of the ribs, so mark off the width of the ribs ($\frac{1}{2}$ in. in this case) and cut away the keel to form a rabbit on each side $\frac{1}{4}$ in. wide, leaving $\frac{1}{4}$ in. of timber remaining in the centre.   This rabbit should not be cut too deep at the first, but only roughly to size, carrying it along the stem as shown on the drawing Fig. 94 at K.   At the stern,

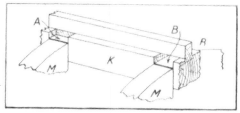

Fig. 95.   Joggling the Keel.

where the deadwood or solid part of the keel must be left thick for the reception of the propeller shaft, the rabbit should be little more than a groove.   Now, as in place of each mould it is necessary to fit a rib, a pocket or joggle, Fig. 95, must be cut to accommodate the rib and to provide a secure fixing, but at the midship sections these pockets may be pierced right through the keel, thus allowing the rib to pass through the hole and making an unbroken rib from gunwale to gunwale, thus considerably strengthening the hull.

This latter course can, however, only be adopted where the cross sections are practically flat. Before the ribs can actually be fitted it is best to secure the gunwales in place. These should be, in the case of the model under consideration, $\frac{3}{8}$ in. square, and made of ash

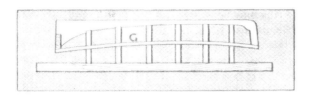

Fig. 96. Fitting the Gunwale.

or oak, and are fitted by marking the sheer of the boat on each separate mould, and cutting a notch $\frac{3}{8}$ in. wide and $\frac{1}{4}$ in. deep into the mould. When all the notches have been cut, a strip of timber, G, may be laid in place, as shown in Fig. 96, for the reception of the ribs. The gunwale is secured at the bow of the boat by recessing it into the stem, securing the same with a small countersunk brass screw.

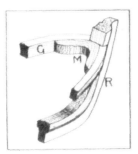 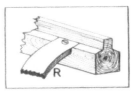

Fig. 97. Securing Gunwale to Stem.   Fig. 98. Screwing Rib to Keel.

To further strengthen the joint, an angle piece of hard wood, M, is cut and fitted in place as shown in Fig. 97, the two gunwales being shown at G, the keel at K, the rabbit cut in the stem at R. The next process is to fit the ribs in position, and this is accomplished by taking a sufficient number of pieces of American elm of suitable length and width and steaming them ; for this a domestic fish kettle may be used, or, failing this, a large pot may be placed over a fire, a piece of wire

netting being hung from the top, and above the water level, so that
the timbers are kept in the steam; after sufficient steaming, the timbers will become quite soft and pliable.   When they are in this state,
select a piece sufficiently long to suit the midship section.   The rib
is first passed through the keel piece, and a copper nail driven partly
through the same to secure the rib from moving.   It is then bent
downwards equally on both sides until it lies fair on the mould.   The
ends of the ribs may be secured, if occasion requires, with a piece of
string tightened up with a tourniquet.   Two copper nails or brass
screws are driven into the gunwales to keep the same in place.   All
the other ribs are fitted in similar manner, except that those forward
and aft must first be inserted into the pocket cut into the keel and

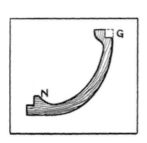

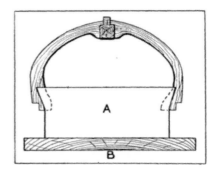

Fig. 99.   A Sawn Rib.   Fig. 100.   Erecting Sawn Ribs an l Keel.

be screwed in place with small brass countersunk screws, as shown
in Fig. 98.
    If sawn ribs are used instead of bent ones, a suitable piece of
elm or oak, preferably with a curved grain, should be selected,
and about $\frac{3}{8}$ in. thick for a metre boat.   Each section has then
to be marked out on the timber to shape as in Fig. 99.   It will be
noticed that the rib is cut to form a knee, N, for the reception
of a screw which secures the same to the keel.   At the upper end of
the edge a notch at G is cut for the reception of the gunwale, and the
rib is strengthened by making it project at that end as in the illustration.   To keep the same in position a temporary cross bar of any
rough stuff is fitted as shown at A, Fig. 100.   This will be secured by a
brad to the base board B.   The planking is accomplished in much the
same way, and this method of construction has much to recommend

it, being quick and simple, but, on the other hand, is somewhat heavier and not quite so strong as the method of using bent timbers. The model is planked under either system in precisely the same manner. Of course, when sawn timbers are used no moulds are necessary : simply fit the ribs, then screw the temporary cross battens or deck beams in place, and fit the gunwales and stringers when the planking may be taken in hand. Fig. 106 shows progress of work at this stage.

It is almost impossible to describe in letterpress the method of cutting and fitting the planks to the ribs, but illustrations Figs. 101 and 102 show two stages of this work. In Fig. 101 the moulds and ribs are visible and the stringers are shown fitted in place. These are continuous lengths of light timber, the width depending on the shape of the hull, but usually about ¼ in. or so wide. The material for the

Fig. 101. The Ribs, Keel, and Stringers in Position.

planking should *not* be cut into parallel strips at the start, as if this is done it will be found that a considerable amount must be cut away to make the plank the right shape. This is shown in Fig. 102 where the plain piece of timber T is shown roughly tacked into place, while the compasses C are being used to "scribe off" the proper shape to cut the plank as shown by the dotted line. The greatest care must be taken to make the joints fit properly. Not the slightest crack between the two planks can be allowed, or the boat will leak. A little experiment will show quicker than pages of reading matter just how to fit the planks. Of course, when one plank is fitted to one side of the boat, use it as a template for the similar plank on the opposite side.

The first plank to be fitted is the "garboard," or that one next to the keel, then the sheer strake, which fits over the gunwale, then alternately, until only one more is required, this is called the filling plank, and requires most careful fitting to insure a sound joint.

There are two methods of securing the planks to the ribs : the first is merely to screw them to the ribs with small brass screws, but this is laborious and requires care to prevent the material splitting. A better plan is to follow the practice of the ship yard, and use copper nails and rooves. Copper nails are of two patterns, one type circular and the others square, and known as rivets. The method of their use is as follows :—The plain nail, as shown in Fig. 103, is first driven right through the plank P and rib R, and of course projects

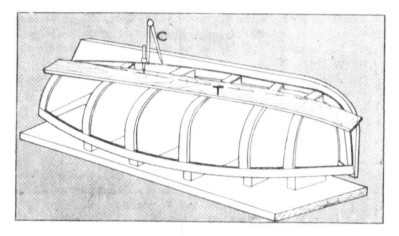

Fig. 102.   Marking Out the Garboard Strake.

beyond the rib, as at A. The head of the nail is then struck sharply with a light hammer turning the point as at B. This process is then repeated by hitting the nail more closely to the rib, finally clinching it right over, as shown at C. While these operations are being carried out it is, of course, necessary to hold a heavy hammer or " dolly " against the head of the nail or the rib as may be, otherwise the nail will come out again, and bend over. Fig. 104 shows the method of using the copper rivet and roove ; the rivet is first driven through the plank and rib and a roove or copper washer, W, is put over the same. A heavy hammer is then held against the head of the rivet and the point hammered up to form a head and so clinch the washer firmly into the rib. To accomplish this nicely, the portion of the nail projecting beyond the washer should not be more than twice its diameter, and any surplus should be cut off with a pair of cutting pliers.   A light hammer will be found most suitable for this work, and once the right knack has been obtained the process can be carried out with great rapidity.

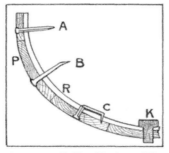

Fig. 103.   Nailing Planks to Ribs.

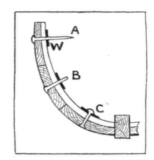

Fig. 104.   Riveting Planks to Rib.

For very large models, intended to carry a crew of one or two men, such as those built in 1913 for the Earl's Court display, a wall-sided, flat bottom construction is advisable, on the score of economy.   The boats in question were on the average 18 feet in length and displaced something like one and a quarter tons.   The hulls were flat-bottomed

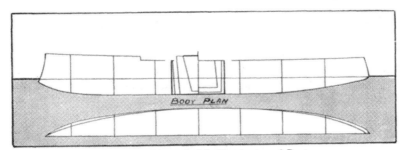

Fig. 105   Lines of 18 ft. Flat-bottomed Boat.

but with a good rise of floor fore and aft, Fig. 105.   The construction was that known as ribband carved, as ribbands or narrow strips of wood were screwed along the inside of the hull, covering each seam.   The illustration Fig. 106 shows two hulls under construction, and makes the system adopted quite clear.

Metal Hulls may be made on somewhat similar lines to those followed in the construction of rib and plank boats, but the ribs are bent from flat metal or preferably angle or tee brass, of suitable section.   The keel is first fashioned of T section brass, with sternpost and deadwood, complete with propeller shaft and stern tube.   To this is soldered and riveted two or three of the ribs, then the gunwales, of flat stuff, are fitted, and the remaining ribs ; two or three stringers or longitudinals are added inside the ribs, deck beams fitted, and the plating is taken in hand.   This is usually of sheet tin or zinc, and

Fig. 106.   A Large Model Ready for Planking.

plates of suitable size are cut, and bent or hammered to shape, soldered or riveted in place, until the whole hull is completed.   No moulds are necessary with this form of construction if every care is taken when building to keep the hull true and free from " winding " or twist. Such methods of construction more or less follow the general style of large shipbuilding practice, but are, of course, much simplified.

The foregoing system is only suited to large models of naval liners or mercantile types.   A more usual form of metal hull is that used for purely sporting or racing purposes. The flat-bottomed, wall-

sided "sharpie" lends itself especially to metal construction. All that
is necessary is to cut two sides and end plate or transom from sheet
zinc. Turn over the edges or fit an angle piece by riveting or soldering, and then fit the bottom and the deck. A tolerably good form
of metal hull construction and largely of one-piece work is made by
cutting a suitable sheet of zinc, tinned steel plate, brass or copper
somewhat on the lines shown in Fig. 107, bending the metal approximately on the dotted lines; the joints at bow and stern are the only
ones that should be soldered and riveted. The result, with a little
care and skill, is a simple form of double wedge displacement

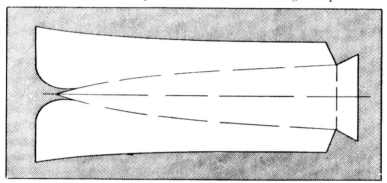

Fig. 107. Setting Out a Metal Sheet for a One-piece Model Hull.

launch, capable of reasonable speeds, and certainly simple and cheap
to make. The form of hull is improved by fashioning the metal on a
curved wood block, of similar lines to the desired hull, but somewhat
smaller to allow of easy withdrawal from the metal skin of the hull.

The deck is cut from one piece of metal, and the coaming bent up,
with a jointing piece added at the forward and after ends to make all
secure and water-tight.

Aluminium does not lend itself readily to model boat work, as
unless one of the special alloys are used the action of the water
speedily sets up a species of decay; and again, aluminium is not
readily worked or easily soldered, although there are some aluminium
solders on the market.

Taking all things into consideration, we are led from personal
experience to the conclusion that a well-made wooden hull on the
"laminated" system is the best for all-round service, because it is
the most accurate to the intended or designed lines, is easily made,

durable, and the interior fittings can be secured in place with a minimum of trouble (Fig. 108).   The weight of the average well-made
metre size motor boat hull is about $1\frac{1}{2}$ lbs. to $1\frac{3}{4}$ lbs.   They can be
made lighter but are then fragile; but in a purely record-breaking
boat, a single-step metre hydroplane wooden hull can be made to
weigh not much over $\frac{1}{2}$ lb. with two coats of varnish.

Other forms of hull construction sometimes seen are paper hulls,
made by gluing pieces of paper at random on a wooden mould of
the desired shape—the mould being thoroughly well coated with

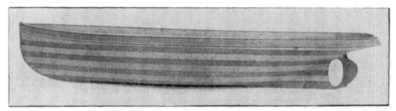

Fig. 108.   A Laminated Hull in Alternate Layers of Pine and Mahogany.

French chalk to prevent the paper sticking to it.   Each layer of
paper is smoothed down before the next is applied, and when finished
the result is a strong serviceable hull, but not very successful if
used to any great extent.   It provides a cheap hull, however, for a
large " spectacular " model that is only required occasionally to
battle with the waves !

Thin veneer as used in some classes of furniture, if well steamed,
makes a similar hull but of better quality.

The author has heard of hulls made of cement, plaster of Paris and
such compounds, but these are scarcely worthy of serious consideration for working models, as, after all, the good old-fashioned wood hull
takes a tremendous lot of beating !

A " Sharpie " Hull in Course of Construction.

CHAPTER VII.

STEAM MACHINERY FOR MODEL POWER BOATS.

THE simplest of all steam mechanisms suitable for propelling a model boat is the very old idea illustrated in Fig. 109, which shows a simple little hull with a curved pipe as shown at A, a spirit lamp being arranged as indicated.   The principle of its operation is simple : the heating of the tube causes the air to be driven out of the pipe, the water flowing in to take its place causing a circulation of water which has the effect of driving the model.   This method, however interesting as a novelty, can hardly be considered as a serious motive power,

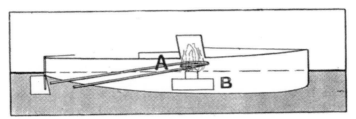

Fig. 109.   Simple Steam Circulator.

and the simplest practical engine is that known as the " Oscillating." Such engines provide an inexpensive propulsive mechanism for a model boat.   A single-acting cylinder can be purchased finished complete for about 9d., while the best quality double-acting engines in the smallest size would cost only about 4s.   The internal arrangements of these engines are very simple, and reference to Fig. 110 will show the whole principle of their operation.   It will be observed that in the single-acting engine, the steam is admitted through the inlet port, and exhausts on the upstroke through the opposite port, the oscillating valve face alternately opening these ports to steam and to exhaust, the steam inlet being indicated at A, and the exhaust at B.   To reverse this engine it is merely necessary to reverse the arrangement of pipes by putting the steam pipe on the port which was previously the exhaust. The double-acting cylinder works on exactly the same principle, but similar ports are drilled at the bottom of the cylinder, as shown at C, Fig. 111. It will also be noticed that the piston rod works through a packed gland, G. This class of engine is well represented by the Stuart Turner " Progress " sets. These little engines are quite useful, with low pressure boilers up to pressures of about 25 lbs., but are hardly suitable for higher pressures as the valve faces are liable to be blown apart by the pressure of the steam ; but for small boats, paddle steamers, and such like craft, these engines are quite reliable.

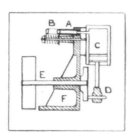

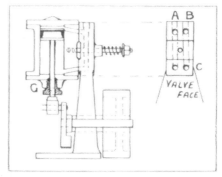

Fig. 110.   Simple-acting   Fig. 111.   Part Section of a Double-acting
Oscillating Engine.   Oscillating Engine.

For those who prefer a more substantial job, cheap and reliable little models are the old-fashioned but none the less useful oscillating engines shown in Fig. 112. These provide an inexpensive but reliable steam engine for a small boat, and, of course, are only suitable for a comparatively slow speed. Two sizes are made as stock lines, and their cost is very moderate.

The engines are made throughout in gunmetal, with non-rusting German-silver piston rods, and although light in design they have ample wearing surfaces. Perfect contact of valve faces is provided by the spring pivot, and the engines are capable of working at speed continuously. Another point in their favour is that they do not take up much head room. No. 1 has cylinder of $\frac{3}{8}$ in. bore by $\frac{1}{2}$ in. stroke, height $2\frac{1}{4}$ in., weight 5 ozs. No. 2 has cylinder of

¾ in. bore by ¾ in. stroke, height 3½ in., weight 15 ozs. No. 1 is suitable for boats up to 30 in., and No. 2 for boats up to 42 in. in length.

For paddle boats it is usual and advisable to adopt two cylinders, each of the double-action type. These can be arranged as shown in Fig. 113, with the cylinders below the crank shaft, which is

Fig. 112. Small Oscillating Engine.

made specially long to carry the paddle wheels. The cylinders only are obtainable in the following sizes :—

No.	1.	$\frac{7}{16}$ in. bore.	1 in. stroke.
,,	2.	½ ,, ,,	1 ,, ,,
,,	3.	⅝ ,, ,,	1 ,, ,,
,,	4.	¾ ,, ,,	1⅛ ,, ,,
,,	5.	⅞ ,, ,,	1⅛ ,, ,,
,,	6.	1 ,, ,,	1¼ ,, ,,
,,	7.	1¼ ,, ,,	1½ ,, ,,
,,	8.	1½ ,, ,,	1¾ ,, ,,
,,	9.	1¾ ,, ,,	2¼ ,, ,,
,,	10.	2 ,, ,,	2½ ,, ,,

The double-throw crank shafts to suit these are illustrated in Fig. 114.

To obtain the best results from a single cylinder engine it should undoubtedly be of the double-acting type, and the Scotch or slide-crank engine, Fig. 115, introduced four or five years ago by Bassett-Lowke, Ltd., is a very successful solution of the problem of designing a powerful double-acting vertical engine that should be comparatively low in height and weight. There are two advantages

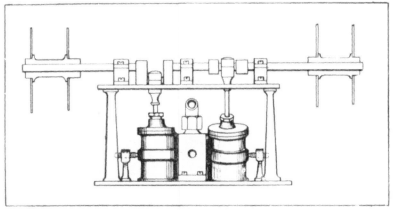

Fig. 113.   Oscillating Paddle Boat Engines.

in reducing the height of an engine for launch work, the first being the ease with which the engine can be installed in varying types of model, and further, the centre of gravity of the engine being low assists the stability of the boat into which it is fitted. These little engines are constructed largely of the best brass and gunmetal,

Fig. 114.   Double-throw Marine Engine Crank Shaft.

and are one of the best engines that can be obtained for model boat work. For steam yachts, tug boats, liners, cross-channel steamers, and such craft, they are without equal, owing to their simplicity and reliability. These engines are made in five sizes, as follows :—

No. 1. Cylinder $\frac{5}{8}$ in. stroke, $\frac{1}{2}$ in. bore, weight 15 ozs. height $3\frac{3}{8}$ in.

,, 2.	,,	$\frac{5}{8}$,,	,,	$\frac{5}{8}$,,	,,	,,	21 ,,	,,	$3\frac{3}{4}$,,
,, 3.	,,	$\frac{3}{4}$,,	,,	$\frac{3}{4}$,,	,,	,,	27 ,,	,,	$4\frac{1}{8}$,,
,, 4.	,,	$\frac{7}{8}$,,	··	$\frac{7}{8}$,,	,,	,,	34 ,,	,,	$4\frac{1}{2}$,,
,, 5.	,,	1 ,,	,,	1 ,,	,,	,,	40 ,,	,,	$4\frac{7}{8}$,,

The disposition of the principal parts of a simple double-action slide-valve launch engine is shown in Fig. 116. The cylinder is shown at A, and as most of the other engines to be described later work on the same lines, a more detailed description of a slide-valve cylinder may with advantage be included here. The two steam ports from the top and bottom of cylinder are shown at B and C, from which it will be observed that they are led to the valve face D. The valve face is enclosed by the walls of the steam chest ; and this in turn is covered by the steam chest cover F, which provides a suitable fixing for the steam pipe H.

Fig. 115. Slide-crank Marine Engine.

The slide valve J is a rectangular block with a depression in its working face and provided on the opposite side with a suitable slot and device for operating

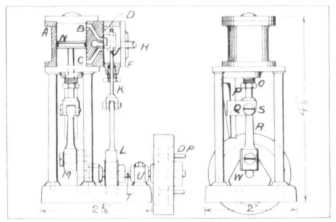

Fig. 116. The Principal Parts of a Launch Engine.

it through walls of steam chest. The valve rod K is alternately pushed upwards and pulled downwards through the agency of the eccentric L, which when properly set relatively to the crank pin M opens the upper valve port to live steam, and at the same time connects the lower port with the exhaust aperture, thus admitting live steam to the top of the piston N, and so driving it downwards, while at the same time the exhaust steam escapes through the port into the exhaust passage and so to the atmosphere. The piston rod works through a steam-tight gland O, which consists, in the simplest models, of a boss drilled and tapped, provided with a small nut, which screws inside it in the manner illustrated. A small quantity of hemp or cotton packing is wound round the piston rod ; the nut is then screwed in place sufficiently tightly to squeeze this packing closely round the piston rod to prevent the escape of steam. A similar packing is also provided in the rim of the piston to ensure a steam-tight joint : on more powerful engines piston rings of

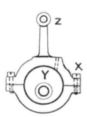

Fig. 117.   Details of an
Eccentric.

Fig. 118.   Set of Parts for " Simplex "
Launch Engine.

springy metal are used. A simple guide bar P, as shown, forms a guide for the crosshead Q, which is screwed on to the bottom of the piston rod. The connecting rod R is attached at the top or " little end " to the crosshead by means of the bolt and nut S, the lower or " big end " in the simplest models being simply drilled to size and made a good fit on the crank pin, but in models with a whole crank shaft, this bottom or " big end," as it is termed, is divided in halves, these being secured together with bolts and nuts. The crank shaft T is carried in suitable adjustable bearings U, mounted on the bedplate, the crank W being made from a piece of mild steel, screwed and riveted into place. This in turn being drilled to the correct distance and tapped to suit the crank pin, the opposite end is counterbalanced to assist in the easy running of the model. The eccentric L and its strap X are illustrated in Fig. 117, which shows one form of this necessary piece of mechanism. The sheave Y has a hole drilled through it out of the centre, the amount of the eccentricity being half the travel of the valve. The strap X, or outer portion, is split into two halves and screwed together, and should be an easy but not a slack fitting on the sheave. The upper end of the rod is arranged to take the small nut and bolt Z, which connect it with the knuckle joint on the bottom of the valve spindle. Two driving pins D P, Fig. 116,

Fig. 119. The "Simplex" Marine Engine.

are provided on the flywheel, and form a flexible connection to the propeller shaft. In the case of the Scotch, or slide-crank engine, the connecting rod is omitted, and a slide crank as illustrated, Fig. 115, is screwed directly to the piston rod, an extension of this rod being carried through the base, to provide it with a steady or guide. The slipper or cross bar has a horizontal slot in which slides a rectangular block of metal drilled with a hole of a size to suit the crank pin. This type of engine is extremely reliable, and one which can confidently be recommended for use on boats where high speed is not the chief consideration. For reliability in operation there are few types of engine to compare with it.

For those who wish to construct their own model launch engine the " Simplex " engine can be obtained as a fully machined set of parts. No lathe work is required for the construction of these engines, as all machining has been done, all drilling and tapping completed, ready for the final fitting and assembling of the engine. The parts supplied comprise everything for the construction of

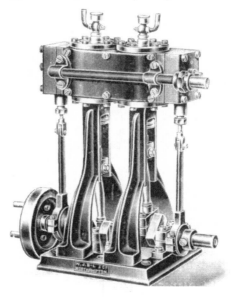

Fig. 120.   Twin-cylinder Launch Engine.

an open column high speed launch engine, $\frac{5}{8}$ in. $\times$ $\frac{5}{8}$ in. All parts are turned, milled, bored and shaped, every hole tapped, and all bolts and nuts, screws and studs are provided, and can be obtained in gunmetal, recommended for use with ordinary boilers, price 10s. 6d. per set, or with cylinder in cast-iron, steel piston, steel crank disc pinned to shaft, and special flywheel, for use with " flash " steam, price 12s. 6d. per set.

A group of these parts is shown in Fig. 118, and the finished engine in Fig. 119. The overall dimensions of the engine are shown diagrammatically in Fig. 116. These sets can be assembled with the minimum of labour, and when finished make a really practical launch engine. Among a host of successful boats fitted with " Simplex " engines may be mentioned the *Lady Alma II*, winner of National Championship Cup, 1913 ; *Alpha*, world's record for her class ; *Nin*, Southern District Champion, 1912, etc.

The desire for greater power in a given space naturally suggests the introduction of multiple-cylinder engines, a standard design of this type of engine being shown in Fig. 120. This consists of two double-acting cylinders mounted on brass standards and bedplate, with a double-throw crank. It is a most sweet running engine,

Fig. 121.   " Stuart " Compound Launch Engine.

and the stock sizes have cylinders with bore of $\frac{3}{4}$ in. by a stroke of $\frac{3}{4}$ in. A further development of twin-cylinder engines is found in the " compound " type. These have one high pressure and one low pressure cylinder. The steam at high pressure enters first the high pressure cylinder, and exhausts thence to the low pressure, where, owing to the greater bore, it is able to perform useful work. Provided the initial steam pressure is high, such engines give good results, but a condenser is practically a necessity. A good example of this class of engine is the well-known " Stuart " small compound shown in Fig. 121. This has high pressure cylinder of $\frac{3}{4}$ in. bore by

$\frac{7}{8}$ in. stroke, and low pressure $1\frac{1}{8}$ in. bore by $\frac{7}{8}$ in. stroke, such an engine being suitable for a liner or large boat about 5 ft. long. The engine weighs about $4\frac{1}{2}$ lbs., and measures some $5\frac{1}{2}$ in. high. Castings or the finished engine can be supplied as desired.

A sturdy engine of somewhat simplified but similar character, originally designed by Mr. J. Carson for flash steam work, proved very successful ; it has cylinders $\frac{3}{4}$ in. and $1\frac{1}{4}$ in. bore, by $\frac{7}{8}$ in. stroke, and is fitted with piston rings. A gear wheel fitted on the shaft is for driving the water feed pump. This engine is now obtainable from Bassett-Lowke, Ltd., either finished, in the rough, or partly machined as desired.

For very high-class working models, into which it is desired to put the very best machinery, the triple expansion engine is

Fig. 122.   The " Stuart " Triple Expansion Engine.

pre-eminent, while in actual service in large vessels it has become almost the standard type. The cost commercially of these engines is too high to admit of their extended use for models, but the castings are obtainable at low price. Fig. 122 shows a particularly neat and pretty design of " Triple " made by Stuart Turner, Ltd. The high pressure cylinder is $\frac{3}{4}$ in. bore, intermediate cylinder. $1\frac{1}{4}$ in., and low pressure $1\frac{3}{4}$ in. bore, with a stroke of 1 in. The amount of hand work involved in the construction of these engines is enormous, but well deserves careful execution, and anyone building one of these engines will be well pleased with the result.

Fig. 123 shows a larger size " Triple " designed by Mr. J. Carson, and now obtainable from Bassett-Lowke, Ltd. This superb model is a reproduction of the engines of Dr. Nansen's *Fram*. Triple expansion, condensing, with cylinders $\frac{3}{4}$ in., $1\frac{1}{4}$ in., and $2\frac{1}{4}$ in. bore, by $1\frac{1}{2}$ in. stroke, it is capable of considerable power. The whole of the motion parts are polished German-silver, the crank shaft being machined from a steel forging, while the bedplate, cylinders, valve chests, eccentric straps, and condenser pumps are of gunmetal. The links are of the solid type, as frequently used in marine work, and the engine is practically an exact copy of its prototype, to a scale of 1 in. to the foot.

Fig. 123. " Triple " Engines of the Model SS. *Fram.*

Quadruple expansion, tandem compound, horizontal high pressure, and similar engines are seldom used for practical working model boats, as their inherent complication, when reproduced in model form, causes serious losses from internal friction. Steam turbines are also not practicable on a small scale, as the successful action of the turbine depends upon the weight and velocity of the steam. It is fairly apparent that although very high *speeds* can be obtained from a small model turbine, no real power is given by this type of engine. It is quite easy to stop a 4 in. diameter turbine by gripping the spindle with a rag held between the fingers.

In purely racing work, the double-cylinder, single-action engine has had a remarkable vogue, and one of the earliest successful commercial models of this class was the well-known No. 1 " Stuart," Fig. 124, made by Stuart Turner, Ltd. This engine has two cylinders, each with a bore of $\frac{3}{4}$ in. and stroke of $\frac{3}{4}$ in.   These engines are now so well known as to need no further description.   There is also a similar engine, the No. 1 *a*, made largely in aluminium, and,

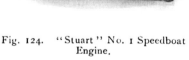

Fig. 124.   " Stuart " No. 1 Speedboat Engine.

Fig. 125.   " Stuart " No. 2 Torpedo Boat Engine for Twin Screws.

of course, much lighter, but only suitable for use with saturated steam.   This engine is intended for boats of light displacement. The whole of the cylinders and crank case is of aluminium, the piston working in a specially hard treble drawn brass liner, pressed in.   The steam chest and cover are of a hard aluminium alloy, and the single slide valve (gunmetal) is worked by a slide-crank on a vertical shaft driven by small bevel gearing ; diameter of cylinder, $1\frac{1}{16}$ in., stroke, $\frac{3}{4}$ in.   These engines cost in the neighbourhood of 30/- finished, and castings may be obtained if desired.

For twin-screw boats the No. 2 engine, Fig. 125, is available.

This has two crank shafts parallel and $1\frac{1}{4}$ in. apart, geared together by two geared flywheels, which ensure both shafts rotating at exactly the same speed. This engine, when supplied with steam at 45 to 50 lbs. pressure, is quite suitable for a metre boat, or a metre and half boat with fine lines.

One of the finest steam racing boat engines yet offered to the public is the "Swift," Fig. 126, designed by Mr. George Winteringham, and built throughout at Northampton Works by Bassett-Lowke, Ltd. This engine is unique in many points ; it is exceptionally light for its power, yet will stand long runs and continuous working under full load for very long periods without distress or signs of

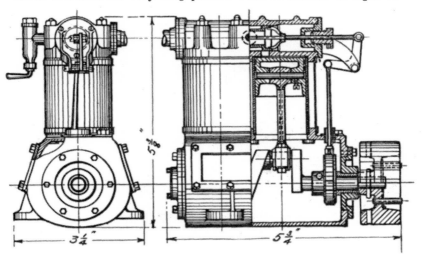

Fig. 126. The "Swift" Racing-boat Engine.

wear, and with the minimum of attention. As will be seen from Fig. 127, it is entirely closed in, runs in oil, thus ensuring effective lubrication. Fitted with cast-iron pistons and cast-iron piston valve of a special mixture, in a cast-iron steam chest, it is eminently suitable for use with superheated steam. The piston valve is balanced, and so arranged that the live steam is in the centre of the valve only, and the exhaust at the ends only. This saves the packing on the valve spindle (the only packed gland in the engine), and also enables a tight packing to be dispensed with, always a source of great loss in small engines, and when working with

superheated steam saves the packing from exposure to the high
temperatures necessary.   There are no losses due to condensation,
as there are no long ports, and the design of the cylinders provides
for efficient lagging of these parts.   The cylinders are ground out,
and the pistons turned to a micrometer fit, and when used with
low pressure of 80 lbs. and downwards, the packing rings may be
entirely dispensed with, thus further reducing the losses by friction.
The lubrication of the valve is attended to by a "Rosco" type
lubricator on the steam chest, and the pistons are lubricated by
the oil in the crank chamber.   For special purposes an extra large
lubricator may be fitted to the steam chest.   This engine can be

Fig 127.   "Swift" High Speed Marine Engine.

thoroughly recommended for boats of 1½ metre, or Class B.
The twin cylinders are 1¼ in. bore, 1¼ in. stroke.   Speed with
saturated steam at 100 lbs. per square inch, 2,200 revs. ; h.p., 1.1. ;
weight without the flywheel, 3 lbs. ; weight of flywheel, 12 ozs.

An entirely new design of " V " engine for speed boat work has
been evolved by Mr. George Winteringham, and possesses many
interesting and unique features.   As may be seen in Fig. 128,
it is a most compact and efficient design, providing large
wearing surfaces to all moving parts, and yet taking up, for
its power, very little room.   It is also very light, and is capable
of being run for long periods without showing any sign of
distress.   The piston and piston valves are very light, and this,

combined with the use of piston valves, and the total absence
of packed glands, makes it a particularly free running engine,
capable of very high speeds. The piston valve is so arranged that
the steam ports are very short, thus reducing steam friction losses
in the ports, a source of great loss in small models. The clearances

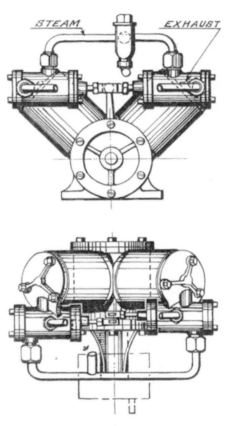

Fig. 128. New Design Winteringham Engine.

are very small indeed, eliminating another source of loss, and there
are no packed glands at all, thus abolishing the greatest source of
power waste in model engines. The piston valve is arranged so
that there is only exhaust steam at practically atmospheric pressure
on the spindle, and the usual packing gland and stuffing box is

replaced by a special long bearing for the spindle, having a bearing surface of 8 diameters in length. This is a perfect fit and shows no leakage. The connecting rods are of special design, giving a truly central thrust on the crank pin, and the shaft bearing is about 5 diameters in length, providing an extraordinary large wearing surface, and reducing oil leakage. The steam ports throughout are

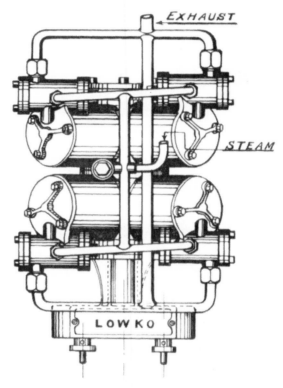

Fig. 129. Four-cylinder " V " Type Racing Engine.

of brass and gunmetal, so there is no rusting up after a run and consequent breakage of some part when starting up again. Provision is made on the steam pipes for lubrication to the piston valves, the pistons and other parts being lubricated by the oil inside the crank chamber. The necessary unions for connecting the steam and exhaust pipes are incorporated in the engine, thus eliminating another source of trouble and expense to the purchaser. This engine is made in four sizes, and can be had in the following arrangements :—

(1) Single engine, twin-cylinder, as Fig. 128.

(2) Double engine, arranged tandem with four cylinders, as Fig. 129.

This design is particularly suitable for use in long narrow boats,

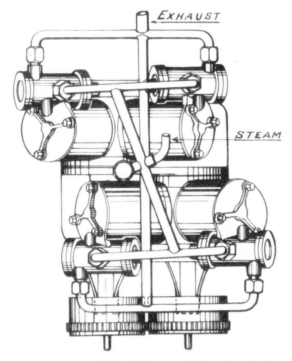

Fig. 130. Four-cylinder Twin-screw Racing Boat Engine.

such as torpedo boat destroyers with twin screws, when fitted with the special gear box shown in Fig. 129, or without this gear if used for single screws.

(3) Double engines arranged side by side with four cylinders, as Fig. 130.

This is also intended for twin-screw boats, the flywheels being geared together.   L

This arrangement of parts combined with the four sizes mentioned is intended to make the design suitable for practically any size boat, or for any sized boiler which may be found most suitable for a particular hull, for it is the boiler which the model can carry which ultimately decides the size of the engine in most cases. If used single or double, the four sizes mentioned above provide many different size power units. For instance, if a boat is long and narrow, too narrow to take a large engine, yet of sufficient length to demand one for full power, a pair of these engines arranged tandem will exactly fill the bill, providing a very narrow engine but plenty of power, there being four cylinders as large as the boiler will steam. The special gear box designed to go with the engine is quite new to model work. First the gearing is all spur gearing, thus obviating the loss of power due to end thrust when bevel gearing is employed. The gears are entirely enclosed in grease, thus providing continual lubrication and keeping the grease where it will do most good ; and lastly the gearing and engine is self contained, thus avoiding the trouble experienced by many when fitting twin screws. The gear wheels being small in diameter permit of keeping the two propeller shafts far apart without unduly raising the centre line of the engine in the boat.

The sizes of these engines are :—

No. 1.—$\frac{1}{2}$ in. bore $\times$ $\frac{1}{2}$ in. stroke.

,, 2.—$\frac{5}{8}$,,   ,,   ,, $\frac{5}{8}$,,   ,,

,, 3.—$\frac{3}{4}$,,   ,,   ,, $\frac{3}{4}$,,   ,,

,, 4.—$\frac{7}{8}$,,   ,,   ,, $\frac{7}{8}$,,   ,,

Each size is made in the following styles :—

Single engine :   2 cylinder with and without gear box.

Tandem engine : 4   ,,   ,,   ,,   ,,

Side by side engine : 4 cylinder without the gear box, but with both shafts geared together.

A very interesting launch engine, shown in Fig. 131, has been constructed by Mr. E. Soulsby of Hull. This is particularly interesting, as it shows the engine, shaft, stern bearing, rudder post, and propeller in their respective places.

To select the best engine to suit a given hull, or for a particular duty, is always difficult. In practice, considerations of expense usually compel the adoption of a standard or commercial type of engine, as the cost of special patterns and castings is considerable, and even then, without experience, the result may not be a success. But there are certain broad lines that can always be followed with success. For example, a tug boat requires a great hauling power at a comparatively slow speed, hence a larger bore engine with lower steam pressure will be better than a high pressure, small bore engine, and internal or engine friction is not so noticeable in this case. For liners, tramp steamers, and slower speed vessels, compound engines have the advantage of moderate speeds of revolution, and economy of steam consumption, with increased length of run from a given boiler.

Fig. 131.   Mr. Soulsby's Marine Engine and Shafting.

For light fast boats, high speed engines of the single-cylinder double-acting, or twin-cylinder single-acting type are imperative with light quick steaming boilers : the natural result of the combination being a faster boat, but with reduced length of run, unless, of course, special provision is made for feed water supply to the boiler.

While for very high speed racers the highest possible engine speeds under running conditions are necessary, the engine, boiler, etc., must be as light as possible. For such conditions a flash steam plant is best, as this type of boiler is the lightest. The water need not be carried in the boat, or the boiler, as it can be pumped in from the pond direct. The engine, however, should be of the double-acting type, with preferably two cylinders. The apparent advantages of the single-acting engine are not so real as they seem, for with a given cylinder capacity the double-acting engine will give more efficient service than the single-acting ; for several reasons the heat radiation is less, there being only one cylinder and one piston in the double-acting engine ; against this must be set the friction of the piston gland and guide bar, but in favour of the double-acting engine, it must be remembered there is only one connecting rod, and only one crank pin instead of two. The balance of advantages,therefore, is in favour of the double-

TABLE No. 7.

ENGINE AND BOILER SIZES FOR STANDARD HULLS.
(BASSETT-LOWKE, LTD.)

Length and Type of Boat.	Engine.	Boiler.
Cargo, 36 in.	No. 782-1 Oscillating	No. 1 W.T.
Cargo, 4 ft. 6 in.	" Stuart " Compound	No. 4 S.F. Launch.
Tug Boat, 4 ft.	No. 4 Slide-crank.	No. 3 S.F. Launch.
Steam Yacht, 33 in.	No. Slide-crank.	No. 1 W.T.
Steam Yacht, 39 in.	No. 2 Slide-crank.	No. 2 W.T.
Steam Yacht, 42 in.	" Simplex."	No. 1 S.F. Launch.
Liner, 78 in.	" Swift " Launch.	" Yarrow " (large size).
Moto Boat (metre).	" Simplex."	No. 1 S.F. Launch.
Sharpie. Class A. Racer.	" Simplex."	Flash 404-1.
River Class T.B.D.	No. 2 " Stuart."	No. 1 S.F. Launch.
T.B.D., 4 ft. 6 in.	Two No. 2 Slide-crank.	No. 2 S.F. Launch.
T.B.D., 5 ft. 6 in.	Two No. 1 " Stuart."	" Yarrow."
" Minotaur," 4 ft. 6 in. Cruiser.	No. 2 " Stuart."	No. 2 S.F. Launch.
" Orion," 3 ft. 7 in. Battleship.	No. 1 Slide-crank.	No. 1 S.F. Launch.
" Swift," 6 ft. 6 in. T.B.D.	Two " Swift."	Special Flash.

W.T. = Water Tube. S.F. = Single Flue.

acting single or twin-cylinder engine, the twin-cylinder giving much better results than a single-cylinder, cubic capacity of both engines being alike : because the torque or turning motion of the twin-cylinder is more even and regular than it is with the single-cylinder ; thus not so much energy is lost per revolution of the propeller, in accelerating and retarding the water column, due to variations in the pressure on the crank pin and the propeller blades.

It is, of course, impossible to give in tabular form the correct size of engine to suit any and every form of hull, but by the courtesy of Messrs. Bassett-Lowke, Ltd., Table No. 7 of standardised engines and hulls of different sizes has been prepared, and is herewith given, as a reliable guide to the size of engine required for different purposes.

When designing a new boat and it is desired to ascertain the approximate size of engine for it, Table No. 8 may be consulted. This has been prepared as accurately as possible, but in small model power boat engines so many apparently trivial details crop up, and modify the expected result, that the table should be used with caution, and accepted only as a *basis* for the design. For instance, two engines may appear to be exactly alike, yet with steam from the same boiler at the same pressure may run at considerably different speeds, due mostly to differences in the accuracy of fit of the piston rings, and internal engine friction.

The method of using Table No. 8 is as follows :—Ascertain from Chapter IV the probable total resistance ; and the suitable power necessary to overcome this resistance ; say this is 27 Watts. Then determine *either* the appromixate propeller speed, or cylinder size. Suppose a propeller speed of 1000 revs. per min. is considered suitable, look under the vertical line of engine speeds marked 1000, until the line marked 27 Watts is found ; it will be seen that the horizontal line is marked $\frac{1}{2}$ in. by $\frac{1}{2}$ in. This indicates that a single-acting engine of $\frac{1}{2}$ in. bore and $\frac{1}{2}$ in. stroke at a steam pressure of 150 lbs. per square inch should develop at 1000 revs. a power equal to 27 Watts. These results must be doubled for a double-acting engine, and also be multiplied by the number of cylinders. A reference to the tables in Chapter VIII, on boilers, will readily show the suitable boiler size. If this proves to be practicable the design can proceed ; if not, it remains to be decided if the propeller speed should be increased, and the size of the engine decreased, or *vice versâ*. A little

TABLE No. 8.

INDICATED POWER OF MARINE ENGINES IN WATTS, AT VARIOUS SPEEDS AND PRESSURES.

Bore and Stroke of Engine in Inches.	REVOLUTIONS PER MINUTE.																							
	250			500			750			1000			1500			2000			3000			4000		
	PRESSURE IN POUNDS PER SQUARE INCH.																							
	25	50	150	25	50	150	25	50	150	25	50	150	50	100	150	50	100	150	50	100	200	150	250	350
½ in. by ½ in.	1	2	6	2	4	13	3	6	19	4	8	26	12	24	38	16	35	52	26	52	104	104	175	245
⅝ in. by ⅝ in.	2	4	12	4	8	26	6	12	38	8	16	52	24	43	70	32	70	104	52	104	208	208	363	500
¾ in. by ¾ in.	4	8	24	8	16	43	12	24	70	16	32	96	47	94	141	64	123	192	96	192	384	384	627	871
⅞ in. by ⅞ in.	6.5	13	39	13	26	78	20	40	120	26	52	156	80	160	240	104	203	312	156	312	624	624	1070	1495
1 in. by 1 in.	9.5	19	57	19	38	114	30	60	180	38	76	228	120	240	360	152	304	456	228	456	912	912	1613	2259
1⅛ in. by 1⅛ in.	13	26	78	26	52	156	40	80	240	52	104	312	160	320	480	208	416	624	312	624	1248	1248	2136	2990
1¼ in. by 1¼ in.	18	36	108	36	72	216	56	113	339	72	144	432	226	452	678	288	5·6	864	432	864	1728	1728	3022	4231

The above table gives the theoretical power developed in one cylinder of a single acting engine of the sizes and at the speeds and pressures given.

For a single-cylinder double-acting engine, the Watts figures must be multiplied by two.

For a double-cylinder single-acting engine, the Watts figures must be multiplied by two.

For a double-cylinder double-acting engine, multiply the Watts figures by four.

Allowance must be made for the loss of power in engine friction—usually a well-made model engine absorbs about 40 per cent. power in friction. Thus, to obtain a workable result the figures in Watts must be multiplied by .6 for an engine that is 60 per cent. efficient.

In the above table no allowances have been made for early cut off, or for condensation losses, etc.

Fig. 132. Pair of Simplex Engines Geared Together for Driving Twin Screws.

study of the tables will prove of help in estimating the general all-round practical size of plant for the power or speed desired.

Naturally these tables cannot be infallible, and all the conditions of each individual design must be thought out in conjunction with one another to insure a good result.

No details of the mechanical processes of launch engine construction are possible in a work of this size, but reference should be made to the excellent *Model Engineer* series of text-books on metal turning, use of tools, and such subjects, or a visit paid to the *Model Engineer* laboratory for practical instruction in model launch engine construction.

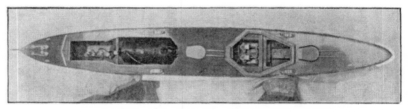

Fig. 133. A Battleship Model Showing Machinery in Position

CHAPTER VIII.

STEAM BOILERS AND BURNERS.

THE boiler may rightly be considered as the heart of a model boat, and as in the human body the heart must be strong and reliable, so with the model boat. The boiler must be, above all things, strong and reliable, capable of constantly and consistently supplying the energy—steam—to the engines.

A brief consideration of the process of steam generation may not be out of place, although space forbids a lengthy disquisition on the subject. If we take a vessel partly filled with cold water, and place it over the flame of a lamp as Fig. 134, the water gradually becomes

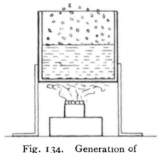

Fig. 134. Generation of Steam.

Fig. 135. Generation of Steam in Enclosed Vessel.

hotter and hotter until it reaches a temperature of 212° F. Beyond this temperature it is impossible to heat water, as at 212° F. it turns into aqueous vapour, or more popularly " steam," the surface of the water being broken up or in "ebullition," a state of affairs known as boiling. Now the steam may be heated to a very high temperature, much in excess of 212° F., and such increases in temperature are associated usually with a rise in the pressure. Suppose now the vessel is entirely closed in, as in Fig. 135, and we heat the water, at 212° F. the water boils, and steam is generated and continues to generate so long as the heat supply is sufficient and undiminished. When the space in the vessel is filled with steam at, say, 1 lb. pressure per square inch, and it is desired to increase the *pressure*, the water must be kept boiling until another amount of steam equal in volume has been generated. Thus there is now twice as much steam in the boiler space, and obviously it is compressed to twice its original pressure, that is, the pressure gauge would now read 2 lbs. per square inch ; similarly, at 10 lbs. pressure, ten times the volume of steam at 1 lb. pressure must be generated and packed into the space in the boiler, and so on. At 100 lbs. pressure, therefore, there is one hundred times the volume of steam in the boiler there was at 1 lb. pressure.

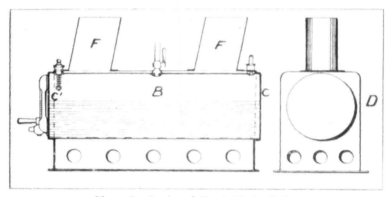

Fig. 136.   Section of Simple Marine Boiler.

To maintain the supply of steam, therefore, at 100 lbs. pressure, it is obvious the source of heat must either be one hundred times the volume required at 1 lb. or one hundred times as hot. Thus the " heating surface " is the governing factor in determining the capacity of a boiler. This should show the enormous demands made on the boiler by a small engine, working at a high speed and high pressure. Consequently, boiler design for power boats possesses much fascination, and calls forth all the skill and ingenuity of the designer and builder.

Fortunately, for ordinary and high speeds, the boilers can be constructed without any very great difficulty, provided suitable materials are used, and all joints well riveted or properly brazed.

This chapter only deals with those boilers that contain their own water supply, "flash" boilers being described in the next chapter.

The simplest forms of model steam boats are usually provided with what is known as an externally fired boiler, Fig. 136. This consists of a plain tube of thin brass or copper B, with two ends C C¹, securely attached to the same. This is enclosed in an outer casing D, of sheet metal, generally Russian iron, and a funnel or funnels F are provided for the dispersion of the fumes or products of combustion.

Firing is effected by means of a methylated spirit lamp which is one of two types, either the older style of plain wick lamp consisting of a reservoir, a central supply tube, with vertical tubes containing the wicks ; or the vaporizing spirit type as shown in Fig. 137. These lamps are provided with a reservoir A for containing the methylated spirit and a small tube B leads a supply of spirit from the bottom of this container to the bottom of the capil-

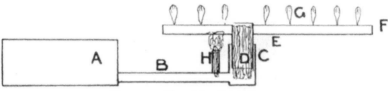

Fig. 137.   Vaporizing Spirit Burner.

lary tube C, which contains a quantity of specially prepared wick D of a sufficient substance to allow of the spirit rising through the same by capillary action.   Outside this tube is another cover tube E, which fits closely.   In the top of this second tube are fixed two burner tubes F, similar in diameter and of a length suitable to the boiler.   In these burner tubes are pierced a number of very fine gas holes G, a small pilot burner H being fitted to enable these gas tubes to be warmed ; the action of the burner being to turn the liquid methylated spirit into vapour, which, mixing with the air at the burner holes when ignited, burns with a clear blue flame.   This type of burner is quite 20 per cent. more efficient than the plain wick burners, thus giving greater steam raising powers to the boiler, as well as increasing the length of time for which it will burn without refilling.

The disadvantage of the plain Pot type of boiler is its low pressure (about 25 lbs. per square inch being the maximum) ; a more powerful boiler of similar design is therefore necessary for higher pressures, and such boilers as shown in Fig. 138 have been evolved and proved satisfactory. The construction of boilers of this type is comparatively simple, but nevertheless requires care to insure satisfactory results. The metal of which the boilers are composed must be of the best, and of light gauge. The ends and water tubes must be properly brazed in, and the whole installed in a sheet metal casing with asbestos lining. Solid drawn copper tubing not less than 18 gauge should be used. Cast gunmetal ends are convenient in use and neater than flanged ends, while the metal is thick enough to take the ordinary fittings by drilling and tapping directly into the end.

A number of water tubes are brazed in the underside of the boiler

Fig. 138.   Simple Water Tube Boiler.

barrel, and of course insure a good circulation of water, which increases the effectiveness of the boiler. This form of construction provides a strong boiler capable of being worked at high pressures of from 35 to 75 lbs. per square inch, with perfect safety. The boiler should be provided with a number of fittings: usually a filler, safety valve, screw down stop valve, for regulating the steam supply to the engine, a water gauge to show the level of water in the boiler, pressure gauge with syphon to show pressure of steam. The syphon may either be a neat brass fitting or a bent copper pipe, the object in both cases being to provide a trap or pocket in which the condensed water may accumulate and keep the hot steam from direct contact with the delicate mechanism of the gauge.

A check or non-return valve is frequently provided, and is used for the attachment of the delivery pipe from a feed pump, the use of this being to allow of a supply of feed water being pumped into the boiler against steam pressure, thus obviating the necessity of withdrawing the lamp and reducing the steam pressure.

The boiler proper is then mounted in a sheet iron or steel case, and the funnels are attached thereto, the object, of course, being to enclose the flame of the lamp, and to protect the boiler from access of cold air.   The great point to be remembered is to provide sufficient air for proper combustion, so that the burner may work properly.   If the air inlets were restricted the lamp would be choked and give off unpleasant fumes ; at the same time its heating qualities would be considerably reduced, and in extreme cases the lamp would either not burn at all, or would beat back, that is to say, the flames instead of going around the boiler would beat out of the fire door into the body of the boat.

Suitable proportions for boilers of this class are given in Table No. 9, and by means of Table No. 7 the approximate size for a particular engine may be selected.

TABLE No. 9.

SIMPLE WATER TUBE BOILERS.

No.	Diameter of Barrel.	Water Tubes Diameter.	No. of Water Tubes.	Casing Sizes.			Approx. Weight Empty.		Approx. Heating Surface in Square ins.
				Length.	Width.	Height.	lb.	oz.	
				ins.	ins.	ins.			
1	$1\frac{3}{4}$	$\frac{3}{16}$	2	$8\frac{1}{2}$	$3\frac{1}{4}$	$3\frac{1}{4}$	2	12	30
2	2	$\frac{3}{16}$	3	10	$3\frac{3}{4}$	4	3	6	57
3	$2\frac{1}{4}$	$\frac{1}{4}$	4	$10\frac{3}{4}$	$4\frac{3}{4}$	$4\frac{3}{4}$	5	12	74
4	3	$\frac{5}{16}$	5	12	6	$6\frac{1}{2}$	7	4	90

This class of simple water tube boiler is only suitable for moderate pressures and comparatively slow speed engines, but has the advantage of long steaming capacity.

Two methods of firing this type of water tube boiler are available ; the first is to use a methylated spirit lamp, employing an asbestos wick and a displacement spirit container. The burner itself, Fig. 139, consists of a shallow pan P, the spirit entering the same through small holes drilled in the supply pipe H.   This supply pipe leads directly into the sump J. A small supply of spirit is fed into this sump from the reservoir R, the supply being automatically regulated by means of the spirit tube B, Fig. 140, and the air tube A. The action is extremely simple : the reservoir is three-quarters filled with spirit and the filling cap F screwed down so that it is air-tight. The valve D on the spirit pipe is then turned on, and the spirit flows into the sump J, and so to burner P until its level rises sufficiently to close the bottom of the air pipe A, when, as no more air can flow into the reservoir through this pipe, the spirit is automatically prevented from flowing through the supply pipe, by the lock caused in the reservoir, and consequently, until the lamp has burned sufficient of the spirit to cause the level to drop, no more spirit will be supplied to the lamp. As soon, however, as the level drops below the bottom of the air pipe, the air tension in the reservoir is released, the spirit flows out, air flows in, until the level of spirit in the sump again rises and again blocks the entry to the air pipe. A modification of

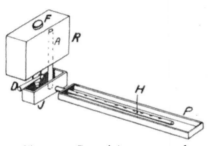

Fig. 139. General Arrangement of Methylated Spirit Burner.

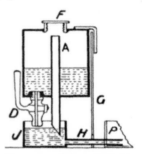

Fig. 140. Automatic Spirit Feed for Burner.

this scheme is found by fixing a container into the boat, and embodying the sump in the spirit container itself; the same operations take place as before, but with this there is no fear of the sump itself catching fire, due to a blow back, but the system has this grave objection, the burner itself cannot readily be withdrawn unless the union is unscrewed.

A different type of boiler and of an improved form is the single-flue launch boiler, Fig. 141, and although there are many competitive boilers put forward as being the best for all-round model work, the single-flue launch boiler undoubtedly holds pride of place for reliability, steady steaming qualities, low centre of gravity, and light weight. For special purposes, of course, other boilers have superior advantages, but for all-round use in fast boats, models of steam yachts, warships, models of the mercantile marine, and so forth, the single-flue launch boiler has very much in its favour. Its circular form is an advantage, as it is easily installed low down in the bottom of the boat, while the absence of a sheet metal casing, exposed to the direct action of the furnace, renders the lagging of the outside of the boiler unnecessary under ordinary circumstances, as an outer shell of a single-flue launch boiler is not exposed to the

Fig. 141. Single-flue Launch Boiler.

direct heat from the lamp, because the heat of the burner flame passes into the furnace tube, where the whole of its heat is used to the best advantage, and radiation losses are reduced to a minimum.

The barrel and furnace tubes of the boilers supplied by Bassett-Lowke, Ltd., are constructed throughout of solid drawn copper tube, special cast gunmetal ends are brazed in, and all joints brazed to ensure strength and reliability. The fittings are reduced to the absolute minimum, and consist of water gauge, pressure gauge, and syphon, screw down stop valve, safety valve ; but for some boats a steam blower is desirable. The dimensions of the stock boilers are given in Table No. 10. Fig. 142 is a diagram of these boilers showing their arrangement and giving the key to the lettering in the table of sizes.

TABLE No. 10.

SIZES OF SINGLE-FLUE LAUNCH BOILERS.

No.	A	B	C	D	E	F	G	Weight, empty.	Heating Surface in square ins.
	in.	in.	in.	in.	in.	in.	in.	lbs. ozs.	
1	8	$3\frac{1}{4}$	$1\frac{3}{4}$	$\frac{5}{8}$	1	$1\frac{1}{4}$	$3\frac{3}{4}$	2 8	46
2	9	4	$2\frac{1}{4}$	1	$1\frac{1}{4}$	$1\frac{3}{8}$	$6\frac{1}{2}$	3 14	65
3	$11\frac{1}{2}$	5	$2\frac{1}{2}$	$1\frac{1}{8}$	$1\frac{1}{2}$	$1\frac{5}{8}$	8	7 4	90
4	13	6	$3\frac{1}{2}$	$1\frac{1}{4}$	$1\frac{3}{4}$	$2\frac{1}{8}$	9	10 12	150

For very fast boats or for use in conjunction with compound or triple expansion engines, the Yarrow type of water tube boiler can be recommended. The principle of this boiler is somewhat different from the preceding types, as one large steam and water drum D is situated at the top of the boiler, two small water drums W at the

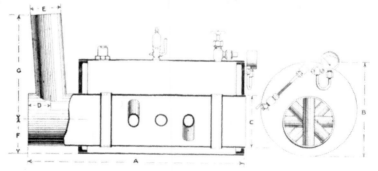

Fig. 142. Diagram of Single-flue Launch Boiler.

lower portion of the boiler, as in Fig 143. These are connected by straight and curved water tubes T, and the furnace is arranged within the arch-shaped cavity so formed, the result being a boiler that will steam very fast, but of course with a limited water range.

The Yarrow boiler may be fired either with solid fuel, provided of course a proper grate is used, or with a methylated spirit lamp, or petrol blow lamp. The casing must necessarily be heavily lagged, and altogether the boiler is likely to come out on the heavy side. A finished boiler and casing is shown in Fig. 144.

For racing boats, where it is not desired to use flash steam, the Scott boiler is to be recommended. This is shown in the photograph Fig. 145, and is a combination of the simple water tube boiler and the semi-flash boiler. It consists, as can be seen by Fig. 146, of a strong steam and water drum D, from the bottom of which depend a number of short water tubes W, arranged as shown in the illustration, thus forming two complete series of coils of water tubes,

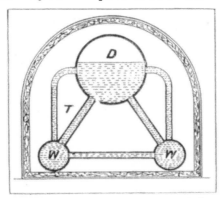

Fig. 143.   End View of Yarrow Type Boiler.

having a large heating surface and causing a very rapid circulation of the water. To obtain dry steam from such a boiler a dome G is essential, and from this dome the steam may be taken direct to the engine. The fault with this boiler is the shortness of time it can be used without replenishing the water. This objection is overcome by a small water feed pump on the engine, or other suitable position, and it is of course essential that the boiler be steamed by a blow lamp of suitable power.

A boiler of this type designed by the author steams the " Swift " engine, Chapter VII, at full speed when fired with a double burner blowlamp of sufficient power.

Table No. 11 gives useful data of these boilers, which can be thoroughly recommended for fast boats.

M

Fig. 144. Yarrow Boiler and Casing.

Fig. 145. Scott Type Marine Boiler.

TABLE No. 11.

DIMENSIONS OF SCOTT TYPE BOILERS.

No.	A	B	C	D	E	F	Heating Surface in square inches.
	in.	in.	in.	in.	in.	in.	
1	2	$3\frac{3}{4}$	$5\frac{1}{4}$	7	$\frac{1}{4}$	4	145
2	$2\frac{3}{4}$	$4\frac{3}{4}$	6	9	$\frac{1}{4}$	$4\frac{1}{2}$	195
3	$3\frac{1}{4}$	$5\frac{1}{2}$	$6\frac{1}{2}$	$10\frac{1}{2}$	$\frac{5}{16}$	$4\frac{3}{4}$	300

The Scotch or return tube type of boiler is sometimes useful in special cases, such as an awkwardly arranged tug boat. This boiler is a combination of the centre flue and locomotive types; it is expen-

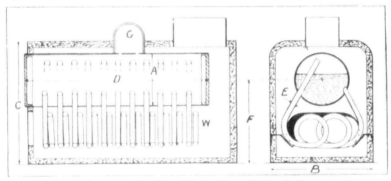

Fig. 146.   Section of Scott Type Marine Boiler.

sive to make, heavy in weight, but a quick steamer and very steady and reliable in use, the general arrangement being shown in Fig. 147, while Table No. 12 gives a range of sizes for such boilers, especially suitable for large models using solid fuel.

TABLE No. 12.

RETURN TUBE BOILERS.

No.	A	B	C	D	E	F	Heating Surface in square inches.
	in.	in.	in.	in.	in.	in.	
1	30	22	15	8	14	1	1525
2	32	24	16	8	17	1	1800
3	36	27	18	9	14	$1\frac{1}{2}$	2485

Locomotive type boilers are seldom used on model boats except of the stern wheel type, while vertical boilers are equally scarce. Both suffer from too high a centre of gravity, and are generally unsuited for model power boat use, except under special circumstances.

As regards the constructional details of the before-mentioned boilers, the smaller sizes are best brazed together, riveting the larger boilers with single or double rows of rivets. Want of space precludes detailed instructions for the mere construction of these boilers, but the case of a single flue launch boiler may be given as showing the general procedure with the smaller boilers, reference being made to the *Model Engineer* handbook, " Model Boiler

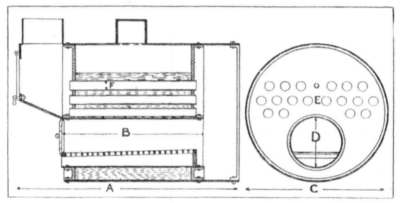

Fig. 147. Section of a Return Tube Launch Boiler.

Making," for further details. Fig. 148 shows a general sectional view of such a boiler.

The flue tube F must be prepared first, and must be marked off and drilled for the cross water tubes T, which may be cut to length, allowing about $\frac{1}{16}$th to project each side beyond the ends of the barrel. The end of this tube is to be clean and bright, and the flue tubes also cleaned. The water tubes are then brazed in place with silver solder, or very soft brass spelter in the usual manner, taking the greatest care to see that each joint is tight and sound. Both ends of the flue tube are then to be trued up in the lathe or with a file, and cleaned up bright. The smoke box end should have a fine thread cut upon the inner surface, employing a chaser

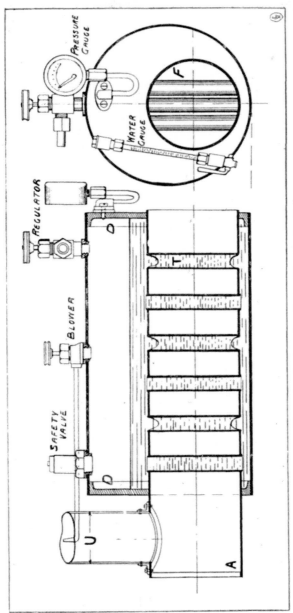

Fig. 148.   Section of a Centre Flue Marine Boiler.

for this purpose. A brass disc is afterwards turned up to correct diameter on the outer edge and screwed to suit the flue tube. At a suitable distance from the end of the smoke box a hole must be cut to accommodate the uptake U, which may be secured in position by means of a small brass angle ring riveted in place. Brazing, of course, must not be used at this point, as the heat of the blow lamp would cause trouble. The uptake is finally fixed in place after the boiler has been brazed together. The outer shell is to be turned up true at both ends or filed correctly to size and the inside polished bright at both ends.

The positions of the bushes for the various fittings must be marked off and silver soldered or brazed in place. The two ends

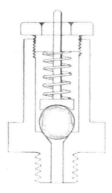

Fig. 149. Section Showing Construction of Perfect Safety Valve.   Fig. 150. Pillar Type Safety Valve.

are turned in the lathe, cleaned up on the outer face and the rim turned a correct close fit in the boiler shell. The two discs D for the ends may then be temporarily soldered together and mounted on the face-plate, the hole for the flue tube bored out, taking care to insure a close fit on the flue tube. The inner face of the boiler ends should be cleaned up slightly at the rim and around the opening for the flue tube, and the whole assembled in readiness for brazing in the ends. This, of course, is the most difficult part of all the processes, and the utmost care must be taken to see that a strong, sound job is made of it.

When this has been finished the fittings may be temporarily screwed in place and the check valve connected to a small force

pump. The boiler having previously been filled with water, a hand
pump is used to pump more water into the boiler, with the result
that the pressure in the boiler will rise. A working pressure of
75 lbs. is usually sufficient, and if the boiler is tested with cold
water to a pressure of 150 lbs. and remains tight, all will be well.
While testing operations are in progress, the boiler should be wiped
perfectly dry, and of course the appearance of even a small drop
of water indicates that one of the joints is not perfect, and this will
have to be remedied either by re-brazing or possibly by sweating
with soft solder, if the leak is not considerable. The pressure gauge,
water gauge and other fittings are best obtained ready for use ; they
can be made, but require special appliances, and are cheap to buy.

Fig. 151.   Plain Plug Cock.   Fig. 152.   Screw-down Steam Valve.

The essential fittings on any small model boiler are a safety valve,
removable to provide means of entry for the water, and a stop valve
or regulator for the supply of steam to the engine.   Those that are
necessary on quick steaming or high pressure boilers are the above,
with the addition of a water gauge, to show the amount of water
in the boiler, a pressure gauge and syphon to indicate the pressure
of the steam in the boiler, check valve for the feed pump connections,
test cock to show level of water, and blow off cock to allow of any
remaining water being emptied from the boiler.

Naturally, the size and arrangement of these fittings vary with
the different boilers, and others may be added to certain boilers for
special reasons, but the fittings mentioned above are those usually
provided and most generally useful.

The most important fitting is the safety valve. This should act surely and with certainty under all conditions, and the type known as the "Perfect" is probably the most reliable valve in a small size that is obtainable. Its construction can be seen from Fig. 149, which is a section of one of these valves, where it will be seen that the valve consists of a steel ball, pressed on to a seating by the plunger and spring, the tension being variable by slackening or tightening the upper nut. The hexagon flange is provided so that a spanner may

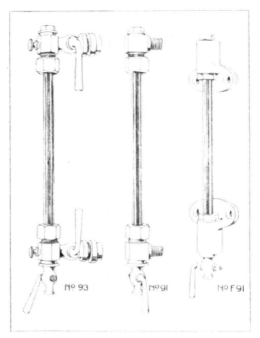

Fig. 153. Water Gauges.

be used to screw the valve in place, a leather, asbestos or metal washer being used to make a steam-tight joint.

Another efficient safety valve is the "Pillar" type, shown in Fig. 150.

Next in order of importance is the steam regulator. This may be either a simple plug cock, as Fig. 151, or preferably a screw-down needle valve, as Fig. 152. These are preferable as they hold high steam pressures without leaks or "weeping." The water gauges,

Fig. 153, are good types; the cocks at top and bottom of main fitting allow steam and water to be turned off if an accident occurs, while the outlet cock at bottom of gauge serves to clear the glass and also acts as a blow-off. A pressure gauge is practically indispensable ; Fig. 154 shows this little fitment.

A check valve as in Fig. 155 is a useful fitting and permits the water from the feed pump to pass into the boiler, but the non-return valve prevents it from being driven back out of the boiler when steam is up.

Fig. 154. Small Pressure Gauge.

Fig. 155. Plain Check Valve.

The launch boilers, Scott and Yarrow type boilers, described on previous pages, are most efficient when fired by means of a petrol blow lamp, Fig. 156. This piece of apparatus is really very simple, and when care is used is quite safe and reliable, while it gives a tremendous heat for its weight and size.

Provided reasonable care and attention are paid to the burner no danger of fire or explosion need be anticipated. The principle on which the burner operates is very simple ; a quantity of petrol is contained in a reservoir, which usually consists of a seamless metal tube with pressed ends brazed securely in position. The reservoir is provided with a non-return air valve, or in some cases with an air pump, and in larger burners a pressure gauge is fitted. From the lower portion of the reservoir a pipe leads up to a screw-down valve provided for the regulation of the flow of the petrol. This valve is in direct communication with the burner, which consists of a coil of copper pipe of small bore, called the vaporizing pipe, which encircles the burner tube, this being provided with air inlets. The vaporizing pipe finally ends in a small nipple situate in the centre and at the end of the burner tube. The method of operating the burner is, first of all to heat the vaporizing coil and burner tube to a dull red heat; this is best accomplished by putting a quantity of asbestos wick in a tray or pan, with some methylated spirits, and igniting the same. In the meantime a small air pressure is pumped up into

Fig. 156.   Simple Petrol Blowlamp.

the reservoir, this causing (when the valve is partially opened) the spirit to flow through the vaporizing tube, where it is turned from its liquid state into a gas, or is vaporized. It then emerges from the nipple in the form of a very fine spray, and induces a draught of air through the air holes, mixing with the air in the burner tube, and finally igniting at the mouth of the burner, where it burns with a hot blue flame accompanied by a loud roaring noise. A little experiment will show the best pressure required in the container, but a pressure of from 10 to 15 lbs. will usually be found sufficient. From time to time the nipples will require cleaning, and the hole should be cleaned with a proper burner pricker ; on no account should a needle or such like be used, as these are far too large for the purpose. The fault usually experienced with handling this class of burner is that the burner is not heated sufficiently before turning on the spirit and that the spirit stop valve is opened too much, thus allowing too great a quantity of spirit to pass, the result being that it burns with a smoky yellow flame, quite useless for effective heating purposes. When this class of burner is used to fire the water tube boilers previously described, it is necessary to cut away the front of the casing, and to have the same lined with asbestos, to prevent the

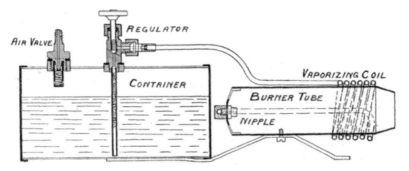

Fig. 157.   The Component Parts of a Blowlamp.

greater local heat with this class of burner from damaging the hull and the boat.

The construction of such a blowlamp is fairly simple ; but skill is required to properly proportion the bore of the nipple, to determine the diameter and length of the burner tube and the area of air space around the nipple.

The dimensions of the Lowko blowlamps as designed by the author are given in Table No. 13, and may serve as a useful guide in preparing a new blowlamp.

Fig. 157 indicates the general internal arrangements of a blow lamp, while Fig. 158 shows a Duplex burner.

The container should be made from seamless brass or copper tube ; the " mandrel drawn " variety will be strong enough for ordinary use.   The two ends can be made by spinning over the edges of a suitable brass disc, or can be purchased ready for use, but they must fit

TABLE No. 13.

Standard Dimensions of Blowlamps.

No.	Container.		Total Length	Burner Tube.		Vap. Tube.			Approx. Weight.
	Dia.	Length		Length	Dia.	Length		Dia.	
	in.	in.	in.	in.	in.	ft.	in.	in.	lb.
1	2	3	5½	3¼	1	2	4	$\frac{5}{32}$	1
2	3	4½	8	3½	1¼	2	8	$\frac{3}{16}$	2
3	3½	6	9¼	4	1¼	2	10	$\frac{3}{16}$	4¼

Fig. 158.   A Duplex Blowlamp for Marine Use.

closely on the outside of the container tubing.   Before being brazed
in position the bushes for the combined air valve and filler, and the
regulator, must be silver soldered or brazed in place.   If this has
been done the two ends may be silver soldered or brazed on.   The
container should be tested to see that it is perfectly petrol-tight, and
should also be carefully tested with an air pump to a pressure of
about 40 lbs. per square inch.

The burner tube should be made from light steel tube, as this is more durable than copper. The outer or burner end is to be nozzled down by mounting the tube on a length of hard wood, leaving the nozzle end projecting, which after having been thoroughly annealed may be forced into shape with a spinning tool.

The vaporizing coil should be made of seamless copper tubing, coiled around the burner tube as shown in the illustrations. The nipples can be purchased for a few pence, but the block in which these are screwed must be cut from a solid piece of brass, and the vaporizing tube brazed into same.

The union for connection of vaporizing pipe to the screw-down valve must be brazed on to the tube, but the downcomer pipe from

Fig. 159.   Steam Machinery for a Fast Model Boat.

the regulator to the bottom of the tank only requires soldering in place.

A disc of thin sheet steel or tin fitted at the back of the nipple of suitable size regulates the air supply, to obtain a perfect mixture ; but the area of this can only be determined by actual trial.

When all is correct, the burner should produce a powerful roaring flame, a blue red in colour, and it should not blow out in quite a strong wind, and should burn steadily until the petrol is practically all exhausted.

The great points to bear in mind are to regulate the area of the burner tube, until a perfect mixture of air and petrol vapour is formed in the tube, and the flame should burn just inside the nozzle of the tube, so that a cross wind will not blow the flame out, as it would if the burner tube were not nozzled down, or proportioned so that the flame commenced at the mouth of the tube.

To start the burner, the vaporizing coils must be heated to a dull red heat, before turning on the petrol, as it is imperative for the success of the burner that the petrol is heated sufficiently to vaporize before it issues from the nipple.

The proper proportioning of a boiler to adequately supply steam to a particular engine running at a known speed is, in small model boat work, difficult to accomplish with certainty, due to the many little items that affect any particular case. Thus, two boilers alike in heating surface capacity may not steam alike, even with the same blowlamp, as a little extra thickness in the flue tubes would affect the steaming properties. This is instanced to show the need of caution in using the foregoing tables, but they are based on standard practice for ordinary boilers, and give a good general guide for the size of a boiler.

Fig. 159 is a reproduction of a photograph of the plant for a twin-screw boat with Scott type boiler, twin engines, and petrol blowlamp, and gives a clear indication of their relative size and arrangement, including the boiler feed pump.

A Racing Boat Hull, showing Twin Screws and Rudder Gear.

CHAPTER IX.

THE system of steam generation known as the flash system, whereby a small but continuous supply of water is forced through passages in a thoroughly heated piece of metal, such as a length of steel tube, is popularly considered as modern in origin and application; but in the reign of George IV steam-driven motor omnibuses were plying for hire between London and Paddington, then a country village. These buses were the forerunners of our modern Clarksons, and had flash boilers designed by Hancock & Gurney. Of course they had their defects, but public opposition and the action of the turnpike and road authorities caused these buses to be discontinued. In more recent years, the famous Clarkson, Serpollet and White steam motor cars have sprung up and operate most successfully. All these are using modifications of the flash type of generator, and perform excellent service. For model boat work the great advantage of this system is its lightness, ease of construction, and the high pressures readily obtainable. The great disadvantage, however, is the high temperature usually associated with flash steam, and also its extreme dryness, thus rendering lubrication difficult.

However, the most successful and many of the fastest model power boats are fitted with flash plants, and the results are extraordinary.

Despite the comparative antiquity of the system there is little reliable data suitable for small model work available, and the best results are generally only obtained after much experiment.

The essential principle of a flash plant is to provide a sufficient quantity of adequately heated metal, and through suitable passages to force sufficient water and neither less nor more than that required to generate, practically instantaneously, sufficient steam for a given engine.

The usual, simplest, and a very practical method is to use a sufficient length of weldless steel tubing, heated by a blowlamp or a coal fire, and through this tube to force the requisite supply of water. There are several systems by which this can be accomplished, the simplest being that outlined in Fig. 160. This shows a simple coil of

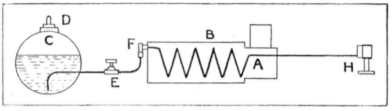

Fig. 160.   Diagram of Air Pressure Feed System of Flash Steam Generation.

steel tube at A, enclosed in a plain metal case or box B, to shield the burner flame from draughts and to reduce radiation losses. The water is carried in a strong metal container C, into which air is pumped through the air valve D, the container being only about half filled with water to allow room for the air. A screw-down regulating valve is shown at E, and this controls the amount of water passing to the boiler, while a check valve F precludes any chance of a blow back

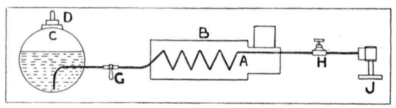

Fig. 161.   Air Pressure Water Feed System for Flash Boilers, with
Steam Regulation.

from the boiler. The water passing, under the air pressure, into the heated metal tube of the boiler is almost instantly turned into steam, and then superheated to a high degree, finally emerging from the end of the boiler tube, and being carried thence to the engine H. It will be noticed that no boiler fittings except the check valve are required, and consequently steam leaks are reduced, weight reduced, and the expense is less, although, in justice to the regular types of boiler, the weight and cost of the water container and fittings thereon should be considered. A modification of this method is to

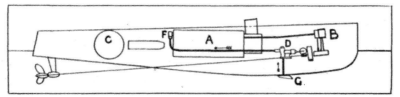

Fig. 162. Direct Pumping System for Flash Steam Generation.

use a plain plug cock G (Fig. 161) for the water supply container, and a screw-down steam valve H on the engine end of the boiler tube. Many consider this an improvement over the water controlled system

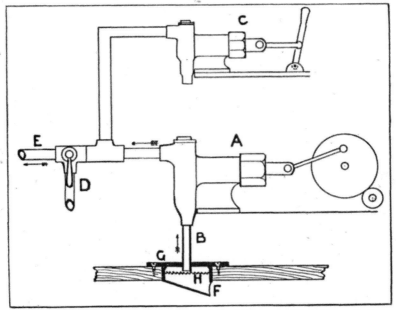

Fig. 163. Diagram Showing Disposition of Feed Pumps or Flash Steam System.

as in Fig. 160, as, of course, when the pressure in the boiler rises above that of the air tank, the remaining water is forced back into the tank, and there being no more water in the boiler, no more steam is generated. **N**

The alternative system is to fit a feed pump or pumps on the engine, and for these to draw their supply of water direct from the pond or a small open top tank carried in the boat, and replenished with water from the pond by means of a scoop or tube carried through the hull of the boat.

This is the lightest and best method, when the pumps are reliable ; otherwise it is best to draw the veil over the attendant troubles ! Every attention should therefore be given to the pumps for this system of boat gear. These are discussed later on in this chapter.

The general arrangement of a direct pumping plant is shown diagrammatically in Fig. 162.

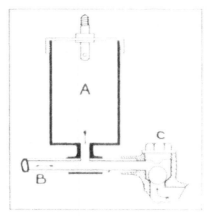

Fig. 164.　Air and Water Vessel for Feed Pump.

The boiler is indicated at A, engine B, blowlamp C, the water feed pump driven off the engine shaft at D, boiler connection at F, and water supply to pump G.

To insure success, however, some further fittings must be ısed, and these are shown (also diagrammatically) in Fig. 163. This indicates the pump at A, suction pipe and strainer at B, separate starting pump at C, and a by-pass pipe and cock at D, the water pipe to boiler being indicated at E. The functions of these various fittings are as follows :—The strainer, shown in detail, is to prevent the entrance of dirt or impurities in the water into the pumps and pipe system, as any such dirt will eventually pass through the valves and engine with disastrous effect. A good form of scoop and strainer is illustrated, and consists of a tube with mouth and hood F, a fixing flange G, and fine gauze strainer H, the water passing up through the pipe B to the pump suction.

A separate starting pump or other suitable device is requisite unless means are provided to turn the engines by hand, thus operating the main pumps and so introducing the water into the heated boiler. It is generally difficult to do this, and consequently a hand pump is usually provided, although a small container A, on the top of the pump delivery valve C, in which water can be introduced and

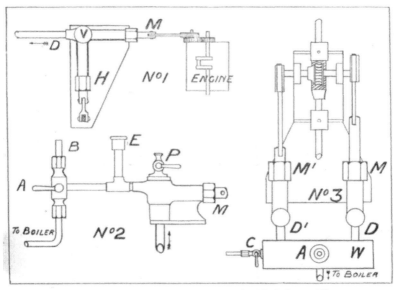

Fig. 165. Sundry Piping Arrangements for Flash Steam Plants.

forced into the boiler B under air pressure, is sometimes used. Mr. Crebbin has adopted this system with success, and it is shown in section in Fig. 164. Another advantage of this method is that the water container serves as an air vessel and pressure regulator when the main pump is working.

The by-pass and cock with overflow pipe, or some such device, is necessary, as any surplus water pumped must be got rid of, and the only thing to do is to pump it overboard. Some means of regulating the amount of water entering the boiler is essential, as if too much is pumped in the coils will cool, and the boiler rapidly

" prime "; or, the contrary, too little water will cause the temperature of the steam to rise abnormally, lubrication of the engine then
becomes practically impossible, and in all probability the engine
will " seize up," if nothing worse happens !

Of various systems of water controlling apparatus some are
shown in Figs. 163 and 165 in diagrammatic form.

Fig. 163 shows the main water pump at A, hand starting pump at
C, and delivery pipes to boiler at E, by-pass pipe and cock at D.
By opening this cock, practically all the water will be pumped
overboard.

In Fig. 165, No. 1, is seen a neat arrangement whereby one set
of pump valves is used and two pump barrels fitted, one operated

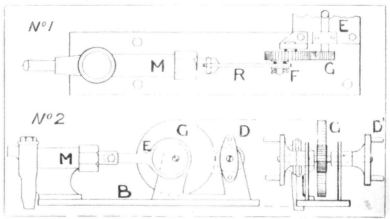

Fig. 166.   Methods of Driving the Water Feed Pump.

by hand for starting purposes, the other driven off the engine
shaft.   This idea was used by Mr. Groves in his racer *Irene*, and
is certainly a very neat and practical device.   The hand pump
must be fixed firmly when not in use, or the ram will work in
and out in unison with the movements of the main pump.

No. 2 indicates a more elaborate control arrangement.   The
pump is shown at M, and from the delivery valve the water pipe
passes to the three-way cock A.   This serves two purposes, the
first to receive, at B, a union connection from a separate small bore
long stroke water pump with flexible metal connection (suitable for
hydraulic work).   This water pump is like a small bore syringe,

and when filled with water is connected to the three-way cock, and
by depressing the plunger of the pump, water is forced into the
boiler, and the engine set in motion, when the main pump takes
up its work.   When the cock is put in the opposite position it opens
this pipe, which then acts as a by-pass pipe, and by this means
serves as a regulator of the water supply from the main pumps.
A " priming " pipe E is fitted to the water inlet pipe, and by
inserting water into this pipe and opening the pet cock P on the

Fig. 167.   A Pair of Double-cylinder Engines, with Pump Gear, for a
6 ft. 6 in. Boat.

delivery side of pump the valves and passages in the pump are
flooded, and it then starts up much quicker.

No. 3 shows the pipe arrangement for duplex pumps.   In their
case it is customary to connect the two delivery pipes D D¹ to a
separate water drum W, taking a single water pipe thence to the
check valve on boiler.   A by-pass cock C and pipe is fitted to
this drum, also an air pressure valve A, or connection for water
pump, to facilitate starting.

There are so many ways of arranging the driving mechanism
for water pumps in the flash steam system of a small boat that

it is almost impossible to describe them all in detail, but some interesting systems are shown in outline in Fig. 166.

No. 1 is a simple arrangement, merely comprising a plain spur wheel reduction driving gear G and a single ram pump M, driven by the eccentric and rod R through the crank-pin F on the gear wheel, the whole mounted on a common base.

No. 2 is a similar arrangement of pump gear, but mounted on an independent baseplate B, and driven by flexible coupling D from the engine shaft, and in turn driving the propeller shaft through another flexible coupling. The great advantage of this system is the facility it affords of detaching the entire pumping gear without disturbing the engine and steam pipes. A development of the same idea is possible having two pumps with operating cranks set at 180 degrees to each other, thus one pump

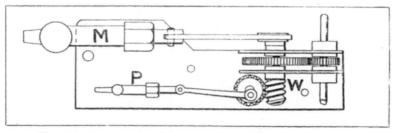

Fig. 168.   Plan of Oil and Water Feed Pumps for Flash Steam Plant.

is forcing water while the other is on the suction stroke, and thus the column of water entering the boiler is very steady, the resulting pressure generated being equally steady. Two pumps of equal size obviously pumping more water than one, it follows that they may be run at a much slower speed, and this consequently is the reason for the worm gear arrangement shown in the figure, and this arrangement can be adopted with every advantage. The disposition of pumps for a very large model boat engine developing over 3 b.h.p. is shown in Fig. 167. In this case the two engines were geared together by a cross shaft and a worm and wheel employed to drive the shaft. Between the two engines, on the end of this shaft, was a crank and slipper, which drove the crosshead, this in turn operating the pump rams. This proved very successful in use.

In Fig. 168 will be noticed, in addition to the water-feed pump M, a small pump P worm driven to run at a slow speed. The worm W is set horizontally to keep the pumps parallel. This is used for the forced lubrication, and it pumps the lubricating oil from a small tank situated in any convenient position above the level of the pump, direct into the valve chest of the engine. Thus a constant supply of lubricant is available for the engine.

Although the author is unaware at present of any practical experiments with such an arrangement, it would appear that an inde

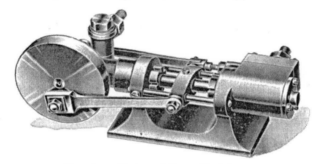

Fig. 169.   Horizontal Steam Feed Pumps.

pendent small steam pump would be very useful for flash boiler work, a pump such as illustrated in Fig. 169 being used. The speed of the pump being controllable by varying the steam pressure by means of a stop valve, it would seem that a practical system could be devised, obviating the need for gears, and greatly reducing frictional losses in the main engine, thus conducing to higher efficiencies. Possibly some readers may try the device. Of course a separate starter would be needed, as with all the other systems.

The construction of a ram pump for boiler feeding is quite a straightforward job, but there are so many admirable examples of commercially produced pumps upon the market at very low prices that it is scarcely worth the time to make the pump itself. The general arrangement of a standard pump is shown in Fig. 170. The pump body B is machined from a casting, the ram R from brass rod slotted at the end A to receive the knuckle pin for the eccentric rod, which is operated by an eccentric or crank. The ram should be a good close fit in the barrel, and is made water tight with hemp packing, secured with gland nut G, which may be fitted with a set screw S to prevent its shifting while in use. The valves are the all-important parts, and plain ball valves are shown in the illustration.

An enlarged view of various valves is given in Fig. 171, No. 1 being the commercial arrangement of ball valve, No. 2 the improved arrangement made by machining the seating to more or less of a knife edge, this providing a gap for the reception of dirt and foreign matter that otherwise would get on to the valve seating and prevent its working. The small cross pin limits the lift and thus prevents the ball from rising too high, and facilitates its rapid closing.

No. 3 is the regulation form of mushroom valve, and is very efficient and reliable in use.

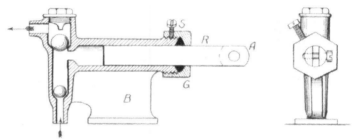

Fig. 170. Section of Ram Pump.

No. 4 shows a spring loaded valve, which is sometimes useful when the valves are sluggish and tend to remain open too long.

Reverting to the general arrangement drawings Figs. 165 and 166 it will be noted that unions should be provided for the suction and delivery pipes, and a screw cap which is fitted over the delivery valve can with advantage be fitted with a pet cock to facilitate the pump's starting.

The speed at which these pumps can be run depends upon many conditions. Originally it was customary to run them at quite slow speeds by worm reduction, but general practice nowadays is to use a $2\frac{1}{2}$ to 3 to 1 reduction gear, and this for a single ram pump is quite all right. For twin or large bore pumps the ratio should be from 8 to 12 to 1, while oil pumps should be geared at least 40 to 1. When quoting these ratios it is assumed that a propeller speed up to 3000 revs. per minute will be used.

The " flash " system has much to commend it, not the least being the fact that flash boilers are very easy to construct, a simple but none the less efficient boiler being sketched in Fig. 172, which shows a plain cylindrical casing, asbestos lined. In this is situate a coil of tube ; the ends are enclosed by light spun metal caps C, and an uptake is arranged at D to carry away the products of combustion.

The tube for the outer casing should be as light as it is possible to obtain, or if it is preferred, the casing may be bent into shape

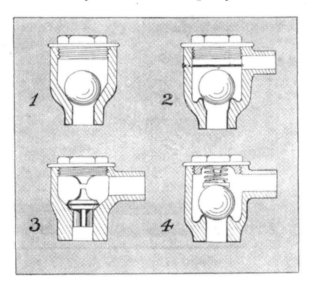

Fig. 171. Types of Pump Valves.

from a sheet of Russian lagging iron, and the seam riveted together. The two end caps should be spun from thin brass or copper, one being merely a retaining cap to cover the asbestos lagging, the other being spun up, as shown in Fig. 172, to form a smoke box. If any difficulty is experienced in making these ends, they can be purchased for a few pence from one of the model boat firms. These ends are spun so that they need driving tightly into place, and will need no further fixing except for a set screw or two put through the lip into the casing. The funnel or uptake consists of a piece of thin brass tubing, and is fitted in place on the smoke box by

cutting a hole through the spun end, flanging out the end of the
funnel tube at either side, and riveting the same securely in place.
The actual boiler consists of, as already explained, a length of
seamless steel tubing.   This should first of all be softened by heating
to a dull red colour, and allowing it to cool slowly.   The tube is
then readily heated over an ordinary gas ring or the domestic gas
stove.   To bend the tube to shape procure a piece of iron bar or
gas barrel, secure this firmly in a strong vice, bend the tube around
the bar, securing one end of the tube to the bench while the other
part of the tube is being bent.

The two ends are employed respectively to take the check valve
Y for the feed water supply, and a union U for the connection to
lead the steam to the engine.   Two holes should be drilled through
the smoke box end cap to allow these pipes to pass through in the

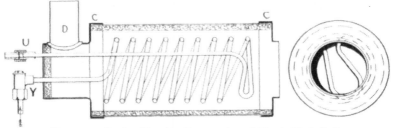

Fig. 172.   Section Showing Construction of Simple Flash Boiler.

manner shown.   Some thin asbestos cardboard should be cut to
the same length as the casing, and wetted, and afterwards inserted
into the casing and secured in place with two metal rings.   The
fixing of the end caps completes the boiler.

The claims made for the success of certain arrangements of the
tubes in a flash boiler are often remarkable, but there is only one
real test and that is efficiency, and this is best obtained by any
method whereby the whole of the boiler tube is exposed directly
to the action of the flames of the blowlamp.

Several typical arrangements are shown in Fig. 173.

No. 1 indicates the arrangement of a plain cylindrical coil of
tube in a rectangular casing.   It will be noted that the coil
is turned upwards into the funnel, to reduce size of casing.

No. 2 is a duplex coil system arranged for use with a single-burner
blowlamp, and shows a compact arrangement of tube.

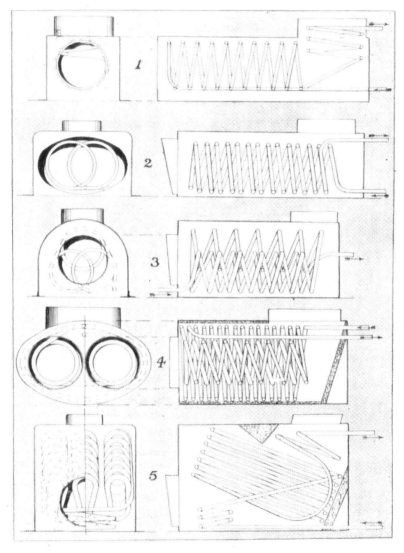

Fig. 173. Examples of Flash Boilers.

No. 3. A triple coil arrangement similar to No. 2.

No. 4. A duplex coil and heating coil system, for use with a double-burner blowlamp. The water enters the outer coil, passes to the centre coils, and through the remaining outer coil to the engine. Such a boiler is shown fitted up in a 5 ft. 6 in. boat in Fig. 174.

No. 5. A straight tube arrangement with an outer coil. A neat and practical arrangement.

Whatever system of tubes is decided upon, the number of acute angle turns should be as few as possible, to reduce the friction of the steam on the walls of the pipe. A very considerable pressure loss is occasioned by mere surface tension and friction between the boiler tubes and the engine.

Fig. 176.   Flash Steam Plant for a 1 Metre Racer.

Unfortunately the amount of data available precludes any definite statement as to the quantities of tube required to drive a given engine at a given speed, but for ordinary purposes it may safely be assumed that 30 ft. of $\frac{5}{16}$ in. tube will generate enough steam for a one brake horse power engine. The average $\frac{5}{8}$ in. by $\frac{5}{8}$ in. high speed engine requires about 10 ft. of $\frac{1}{4}$ in. tube, a $\frac{3}{4}$ in. by $\frac{3}{4}$ in. about 15 ft., while a 1 in. by 1 in. can do with 25 ft. provided it is well arranged and subject to the direct action of the burner flame.

Several examples of high speed flash boiler plants are given in this chapter, and, it is hoped, will be of service to the model power boat enthusiast, in suggesting developments of model machinery construction and design.

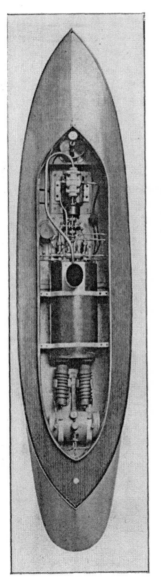

Fig. 174. Plan View of a 5 ft. 6 in. Racing Boat.

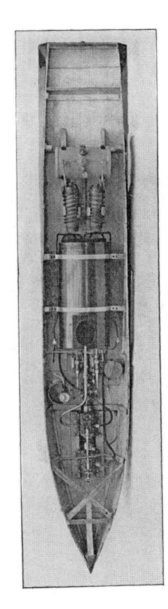

Fig. 175. Plan View of a 1½ Metre Flash Steam Boat.

Fig. 175 is a 1½ metre boat with a four-cylinder engine boiler as No. 4, Fig. 173 ; double-burner blowlamp and duplex pumps somewhat as Fig. 165, No. 4, while a complete plant is shown on the test bench in Fig. 176. This has a "Simplex" engine, duplex pumps, and single coil boiler as No. 1, Fig. 173.

Fig. 177 is a plan view of this plant, which was fitted originally to *Alpha*, and has done much satisfactory work.

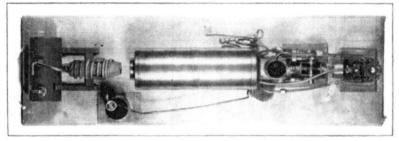

Fig. 177.   Plan View of the Flash Steam Plant in *Alpha*.

CHAPTER X.

THE petrol motor has achieved such remarkable successes on land, driving motor cars, in the air for aeroplane propulsion, and on sea, operating motor boats and commercial vessels, that it is somewhat of a surprise to find that for model racers the petrol motor has only a moderate measure of success in pure speed events. But for fast boats the petrol engine possesses many advantages. The power developed per unit of weight is very considerable, while the engine speed is high but not excessively so. The most modern development of this class of motor is exemplified by the " Stuart " productions, which are eminently reliable and practicable in use, although they are not the only engines on the market ; Messrs. Gamage have a successful engine in the " Stentor," which has won many prizes in open events. For some years past experiments have been made with model petrol motors, but many of the earlier engines suffered from various defects, most of which have now been eradicated. The principle on which the majority of these petrol motors operate is known as the Otto or four cycle, and although somewhat difficult to describe, it is hoped to make the system clear with the aid of diagrams, as a petrol motor of any description is somewhat of a mystery unless the mode of operations is thoroughly understood by the owner. Fig. 178 shows diagrammatically the essential parts of a four cycle petrol motor : these are the cylinder C, piston P, connecting rod R, crank A, flywheel B, timing gears D, valve cam E, exhaust valve F, inlet valve G, sparking plug H, contact maker J, cooling fins K, carburettor L, coil M, accumulator N, switch O. It must be first understood that there is only one working stroke for every two revolutions of the crank shaft, thus the piston rises and falls once without doing any useful driving work, and rises once

again without doing driving work, finally descending for the second
time, when the charge is fired, the force of the explosion driving the
crank shaft round, and sufficient energy being stored into the flywheel to "carry over" the engine till the next firing or working
stroke.   The cycle of operations having four distinct and separate
phases on the first or "suction" stroke the piston descends, by turning the crank shaft by hand, when the vacuum so caused in the
cylinder causes the inlet valve to open, and the rush of air to fill the
cylinder draws a fine cloud of petrol vapour from the carburettor
or "gas works."   The second phase commences with the upward
movement of the piston, known as the "compression" stroke, which

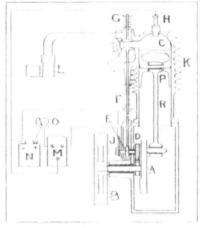

Fig. 178.   Section of Model Petrol Motor.

compresses the petrol vapour gas into a small space at the top of
the cylinder, called the combustion space.   The inlet valve automatically closes, of course, as the piston rises: this stroke is known as
"compression."   The third phase is the working stroke, as the
inflammable gas already compressed in the cylinder is automatically
set on fire, "ignited" as it is termed, by an electric spark, which is
caused to jump across the "spark gap," of the sparking plug at the
instant the piston is at the top of its stroke ; the coil and accumulator,
in conjunction with the contact maker, doing this duty of igniting
the gas, as will be explained later on.   The piston is naturally
forced down violently by the explosion in the cylinder, and this

stroke, known as the " power " or working stroke, *drives* the crank-shaft round. The fourth phase is known as the "exhaust" stroke, as the piston on its upward journey drives the burnt gas out of the cylinder through the exhaust valve to the air, the valve being mechanically opened by the timing gear system at the correct instant. The engine then again commences its cycle of four operations, viz., (1) suction, (2) compression, (3) firing, (4) exhaust. When it is remembered that these separate and distinct functions have to be performed several times per second, while the engine is running, the need for careful adjustment and extreme accuracy becomes apparent

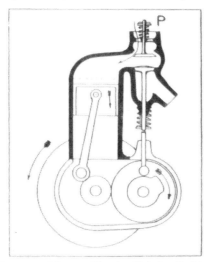

Fig. 179. Diagram Showing Relative Position
of Motor Parts During Suction Stroke.

Taking these four strokes in more detail, reference to Fig. 179 shows the first stroke in the cycle of operations, and represents the piston about to descend, thereby causing a partial vacuum in the cylinder, which causes the inlet valve to open and a supply of petrol or gas to enter the cylinder. The valve shown is of the " automatic " variety; it is normally kept shut by the light spring P, and only opens when the suction in the cylinder overcomes the spring pressure and automatically closes when the compression stroke is about to commence. The mixture is compressed on the upstroke, and immediately the piston reaches the top of its stroke, electrical

O

contact is made, and a spark occurs between the two poles of the sparking plug, thus igniting the gas, which then burns and explodes, driving the piston down on the third or " firing stroke," which is the only power stroke among the four. The fourth remaining stroke completes the cycle of operations by dispersing the burnt gas through the exhaust valve, which in this stroke is opened automatically by means of the half-time gear and exhaust cam, as in Fig. 178. This cycle of operations, known as the " Otto," in memory of the inventor, is the principle upon which most petrol motors operate, but there are others, chiefly that known as the

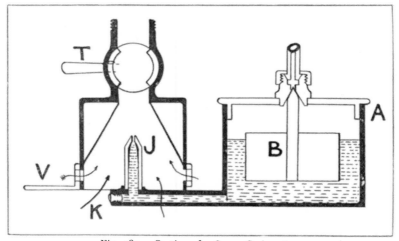

Fig. 180. Section of a Spray Carburettor.

" Two-Stroke," but these are not generally satisfactory in operation for engines so small in size as would be necessary for a model boat, and therefore we may disregard them in this connection.

The supply of gas to the engine is obtained from a piece of apparatus known as the carburettor, and may be likened to a gas works, as its duty is to supply petrol gas from the liquid petrol. A carburettor generally consists of a float chamber A (Fig. 180), in which a light brass or cork float B is arranged upon a central spindle, one end of which is provided with a needle point, the petrol entering through a conical orifice which is automatically closed or unclosed by the needle valve on the float spindle. When commencing to operate the engine the petrol will flow from the petrol tank into the float chamber freely, but as the level of the spirit rises the float will rise with it, until the needle point closes the conical valve, and stops the supply of spirit. Of course, as the spirit is used up, so the float again drops, allowing more spirit to enter, this operation taking place quite automatically, while the engine is running. From the float chamber the petrol is led to the jet chamber, the principal feature of which is

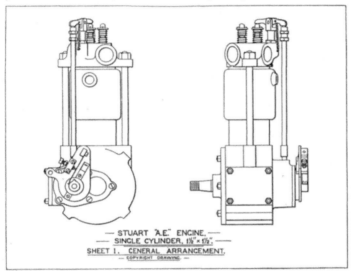

Fig. 181. General Arrangement of "Stuart" Vertical Petrol Motor.

the jet J, and the air inlets K, usually situated at the bottom. It is arranged that the top of the jet is exactly on a level with the highest level of spirit in the float chamber, so as to minimise fluctuations in the supply. The holes in the jet are extremely fine, and on the suction stroke a minute spray of liquid petrol is sucked through this jet, thus mixing with the air which rises through the bottom holes, as shown in the illustration. A throttle T to regulate the amount of gas which passes to the cylinder is usually arranged immediately above the jet, and provision is also made as at V to regulate an additional supply of air to give a more perfect mixture, which, of course, has a direct bearing upon the efficiency of the engine. The illustrations herewith are only diagrammatical representations of the most general arrangements of small petrol motors now upon the market. Sometimes a wick carburettor is used. This is simply a box containing wick soaked in petrol, the vapour given off being drawn into the engine on each suction stroke.

An excellent miniature petrol motor engine is known as the "Stuart" and made by Messrs. Stuart Turner, Ltd., of Henley-on-Thames. The engine is made in four types, two having a single cylinder, and two twin cylinders. The parts in these models are standardised and interchangeable, the only difference being that either the single or twin engine can be supplied for air cooling or water cooling, the air cooled engine being supplied with fins or radiators, allowing the heat to be dispersed very readily. These types are

Fig. 182.   "Stuart" Vertical Petrol Motor.

chiefly used in model aeroplanes or model boats where a very free access and rapid current of air are available. For boat work where the engine is enclosed or screened from the direct passage of the air, it will be found preferable to use water cooling. In this case an aluminium jacket is pressed over the cylinder and a constant stream of cold water is passed through this jacket, a small rotary pump being necessary for this purpose. In a model boat, the water supply can be taken directly through the bottom of the boat, and the outlet passed through the side.

The engines made by Messrs. Stuart Turner, Ltd., are the outcome of lengthy comparative trials of various types. That selected for

the single cylinder is the vertical four cycle model, and embodies
nothing but sound engineering experience.   Fig. 181 shows the
general arrangement of the engine, from which it will be
observed that the valves are arranged in the cylinder head.   A
high ratio of compression is employed   There is an entire absence
of " pockets " in the combustion chamber in which burnt gases
will accumulate, causing fluctuations in power, if not an entire
misfire.   The inlet valve is operated automatically, the spring
being outside and easily accessible.   The exhaust valve is of necessity operated mechanically, through the medium of a cam and vertical

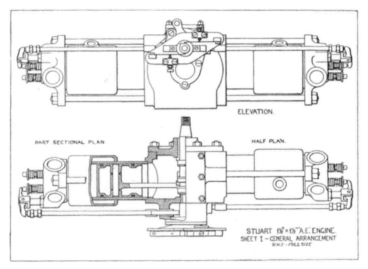

Fig. 183.   General Arrangement of " Stuart " Twin Horizontal Engine

hardened-steel tappet, with hardened adjustable faces operating on
the valve spindle by means of an overhead rocker of hardened steel.
The inlet and exhaust ports are specially large, and provide the
maximum passage for the gases.   The utilisation of a full size
sparking plug insures perfect insulation, and obviates the many
disadvantages of a miniature plug ; in fact, the ignition system
throughout has had the most careful consideration, as so much
depends upon this part of the equipment.   The contact breaker is
of massive construction, and heavily insulated where necessary.

Provision is made by means of the contact screw and lock-nut for accurate adjustment of the contacts, which are iridium faced.

As regards the construction of the engine proper, a novel feature is introduced by the adoption of a specially drawn hard steel tube for the cylinder ; on to this is forced an aluminium water-jacket, or in the case of the air-cooled engine a set of cast-iron fins. The crank case is of aluminium alloy, light and durable, and provided with a large inspection door. Long phosphor bronze bushes are provided for the crank-shaft which is machined from forged steel, and accurately ground and polished. The two-to-one gear for operating the valve is machine cut from cast steel. The connecting rods are of phosphor bronze with adjustable big ends. The pistons are hard cast-iron,

Fig. 184. Twin-cylinder Air-cooled " Stuart " Petrol Motor.

fitted with two piston rings to insure a thoroughly tight fit. The bore of the cylinder is $1\frac{1}{2}$ in. and stroke $1\frac{1}{2}$ in. the power developed being $\frac{1}{4}$ h.p. on the brake. The weight of the engine is $4\frac{1}{2}$ lbs. with the carburettor, which is of the float feed type, and fitted with throttle valve and adjustable extra air valve. The flywheel adds another $1\frac{1}{2}$ lbs., but under certain conditions of design may be omitted for aeroplane or motor-boat work. Fig. 182 gives a good idea of the little engine ready for use. The stand shown is used for test purposes. Provision is made to set the engine in a horizontal position if desired. .

The twin cylinder engine, illustrated in Fig. 183, is of still more interesting design. The cylinders are opposed, and all reciprocating parts are perfectly balanced by the truly scientific balancing of reciprocating masses. The total weight of the motor in comparison to the power is, of course, reduced by using the two cylinders, and amounts to only 8½ lbs. in the case of the air-cooled, and 8 lbs. water-cooled. The flywheel, if used, weighs 1½ lbs. (equal to 16 lbs. per h.p.), a remarkable figure for a soundly constructed practical ½ h.p. engine. The details of construction are identical with the single cylinder model, with slight exceptions. The crank-case is, of course, modified. The crank-shaft is an exceptionally good piece of work, a neat arrangement of the two connecting rods allowing of wide bearing surfaces. The valves and all parts are made to limit gauges, and are all interchangeable. The advantages of the balanced opposed motor for aeroplane work are obvious, while for model racing boats it provides a low centre of gravity, and is readily arranged fore and aft in the boats which may be of quite moderate dimensions.

Fig. 184 shows the twin-cylinder air-cooled model complete and ready for use. It may readily be appreciated that these small motors mark another stride in the construction of sound engineering models.

The ignition frequently provides the cause for a partial breakdown or stoppage of the engine, and some brief description of the principles usually followed may assist to a better understanding of this subject. A supply of low tension electricity is obtained from an accumulator similar to those adopted for model electric boats, but in smaller size ; this low tension current is useless for jumping the "spark gap" in the plug, or overcoming any great resistance, and, therefore, it becomes necessary to increase its tension. This is accomplished by what is known as a "coil," which consists essentially of two separate coils of wire, the primary consisting of a large number of turns of thick wire, and the secondary, S Fig. 185, consisting of a lesser number of turns of thin wire. These two separate

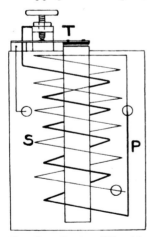

Fig. 185. Section of Trembler Coil.

coils are usually wrapped one over the other and heavily insulated, while in the centre a core of soft iron is fitted, the action of

passing a low tension current in the secondary partially
magnetising the core and causing the blade or trembler T
(Fig. 185) to vibrate, this trembler causing an intermittent current
to flow through the primary P, as it has been found that
this effect produces a more powerful secondary current. This
secondary current is of very high voltage, but very small amperage ;
that is to say, the pressure is very great, but the quantity of current
is very small. The current is, therefore, capable of jumping through
space or of overcoming a heavy resistance or insulation. The
high tension current is led directly from the coil to the sparking

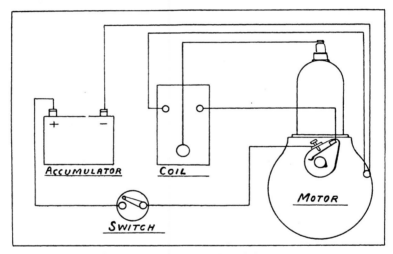

Fig. 186. Wiring for Single-cylinder Engine.

plug, which is simply a terminal for the wire, but owing to the heat
and the high tension current, it is necessary that it be very heavily
insulated, the best porcelain or mica being used. The current flows
through the centre terminal of the plug to the casing, and through
the body of the engine back to the accumulator, thus completing
the circuit, the correct method of wiring to the contact maker
and other parts being indicated in the diagram in Fig. 186. The
contact maker is a simple mechanical device to automatically
make and break the current in the motor, and is shown in Fig. 187.
At A contact is being made, the current from the accumulator
passes through the coil, and the spark takes place at the plug points ;

while at B, the blade C has separated from the contact stud D, as the cam E has revolved, consequently the flow of current has stopped.

To increase the speed of the motor, after it has been started and is running steadily, it is usual to open the throttle, and then to " advance the spark." That is to say, the spark in the engine is arranged to take place slightly before the piston on the compression stroke has reached the top of the cylinder, or rather the top of its travel ; thus the explosion takes place just as the piston is ready to descend on its power stroke, as an appreciable amount of time is required for the gas to catch fire, burn, and expand, or, as it is popularly called, " explode." The timing of the spark is altered by turning the contact maker base bodily around the half

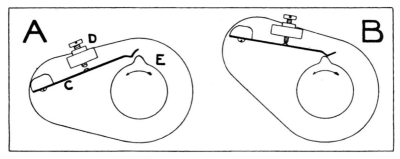

Fig. 187.   Details of Contact Maker.

time spindle, as shown in Fig. 187. At A the firing is arranged for slow running, at B the spark is advanced for high speeds ; it will be noticed that the *blade* has turned towards the oncoming cam, which has therefore pushed it up and made contact earlier, relative to the position of the piston in the cylinder.

The general causes of breakdown or failure of the engine to work will usually be found to be that the points on the sparking plug are either sooty, covered with oil, or even possibly too far apart. The width should not be more than 3 millimetres, or $\frac{1}{32}$ of an inch. The trembler on the coil requires to be kept clean and in good adjustment, so that it may vibrate freely when the current is passing. The accumulator should also be well charged, although a current of 4 volts is invariably adopted. The terminals, and all points of contact between ignition wires and their connections, should be kept clean and well screwed up, otherwise serious loss of current will occur, which will have a very bad effect upon the resulting spark.   When a failure occurs it is best to look out for the following causes for such failure :—

1. See that all connections are properly and strongly made.

2. Examine the sparking plug and see that the points are the correct distance apart, and clean.

3. See that the blade on the contact breaker makes contact with the screw when the engine is turned round, and that these contacts are clean.

4. See that the switch is on, and adjust the trembler on the coil, unless the same is vibrating properly.   This can be ascertained by noticing the buzzing sound which it gives off.

5. See that accumulator is properly charged.

6. See there is petrol in the tank, that it flows to the carburettor, and that the air adjustment is correct.

7. Look at those parts requiring lubrication to see that nothing has seized.

8. Examine the carburettor ; get to ascertain that same is neither choked up nor of too large a bore.

There is no reason why a petrol motor boat should not work properly ; in fact, many very fine performances have been put up by such boats, Mr. Mills' *Stentor Minor* doing about 11 miles per hour.   The general arrangement of a fast boat built and owned by the author is shown in Fig. 188.   The hull of this boat was cut from solid yellow pine, two planks being used for this purpose, the upper one 4 in. thick and the lower 3 in. thick, the joint cemented with marine glue and afterwards stitched with copper wire, the method adopted being to saw out the upper portion of the boat roughly to shape, then temporarily gluing the parts together, and finishing up the hull, both internally and externally, to its approximate dimensions.   The joint is then properly cemented and sewn, the model being finally finished to its correct form and dimensions. Two bulkheads are fitted, the forward one taking the form of a locker, in which are contained the coil and accumulator.   Two propeller shafts, with continuous stern tubes and double " A " brackets, were next fitted, two V-shaped fins of thin sheet brass being fitted in the angle formed by the base of the hull and the stern tube, these pieces materially assisting steadiness of steering.

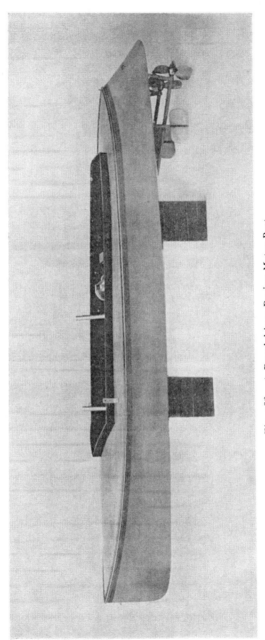

Fig. 188. A Petrol-driven Racing Motor Boat.

Twin rudders were fitted, linked together and operated by a worm and wheel from the interior of the hull, this method providing very fine adjustment, while holding the blades steady and parallel.   The motive power is obtained from a $\frac{1}{2}$ b.h.p. petrol motor having a bore and stroke of 35 millimetres, full power being developed at 1,900 revolutions per minute.   This engine works on the ordinary four cycle, and is fitted with automatic inlet valves and mechanically operated exhaust valves, both being contained in the cylinder head. The cylinders themselves are constructed from a special grade of steel tube, accurately machined and ground out true.   As will be seen from the illustration, the two cylinders are set in the horizontal plane, providing a low centre of gravity.   The carburettor is of the usual float feed type, with throttle and extra air levers, the jet being a single spray.   The contact breaker follows standard practice in

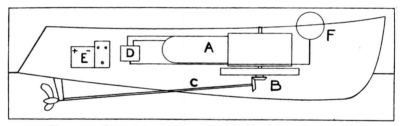

Fig. 190.   General Arrangement Diagram of Petrol Motor Boat.

being the usual make and break type.   The ignition is by twin-cylinder trembler coil, energized by a 4-volt accumulator.   Full sized sparking plugs are used, the wires being heavily insulated and well protected from " shorts."   The transmission is effected by means of a cone clutch contained in the flywheel, and operated by simple cam movement.   From the clutch a light chain transmits the motion to a cross shaft mounted in ball journal bearings ;   on the ends of this cross shaft are two hardened steel mitre wheels, engaging with similar wheels on the ends of the propeller shafts. Ball journal bearings as well as ball thrust bearings are provided for the propeller shafts, the result being an almost frictionless drive, and one of very high efficiency.   The propellers are $4\frac{3}{4}$ in. diameter, 14 in. mean pitch, and each contain 9 sq. in. of blade area.   The lubrication of the model has received special attention : an oil box mounted on two aluminium struts contains an ample supply of

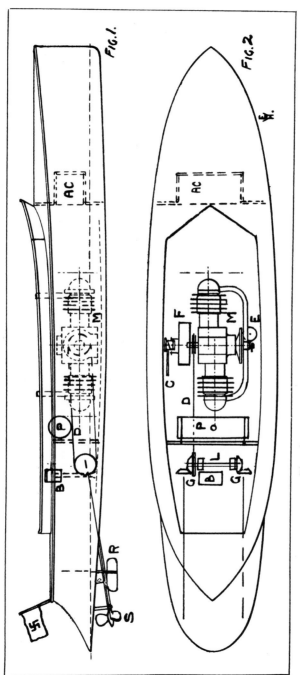

FIG.1.

FIG.2.

Fig. 189.   General Arrangement of a Petrol Racing Boat.

lubricating oil, and this is led through small brass pipes to all the bearings, as well as a separate drip lubrication for the gears and chain, an oil cover or shield of thin aluminium being fitted to prevent oil being scattered in every direction; the petrol tank is attached to a mahogany dashboard in the after part of the vessel, the exhaust pipes being led well past the same. The model is finished in buff, with a blue line, and decked over with the finest yellow pine, provided with a mahogany coaming. The extreme dimensions are 5 ft. 6 in. over all, 14 in. beam, and $1\frac{3}{4}$ in. draught, displacement at rest being 30 lbs.

Fig. 189 is a side elevation in part section, showing the general arrangement of the plant, and also a plan view. Fig. 188 is reproduced from photo of the finished model.

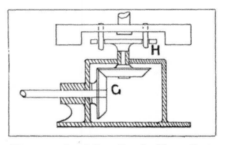

Fig. 191.   Bevel Gear Box for Motor Boat.

The appended list will enable the position of the more important parts to be readily located in Fig. 189.

<blockquote>

A C Coil and accumulator.

B   Oilbox.

C   Clutch lever.

D   Chain.

E   Carburettor.

F   Flywheel.

G   Gears.

L   Cross shaft.

M   Motor.

P   Petrol tank.

R   Rudder.

S   Propellers.

</blockquote>

The arrangement of a single-cylinder engine, as generally adopted by members of the Victoria Model Steamer Club, and others, is shown in Fig. 190. The engine A is set horizontally with the crank-shaft vertical, the upper end has a cross pin for the starting handle to engage with, the lower end is fitted with a simple bevel gear B, driving direct to the propeller shaft C. Carburettor is seen at D, coil and accumulator E, and petrol tank F. An improved driving arrangement adopted by some members is shown in Fig. 191, the bevel gears being mounted in the casting G, the vertical bevel spindle having a pin and slot type of flexible connection H, which facilitates easy withdrawal of the motor for adjustment without the trouble and difficulty of re-aligning the gears.

It is hardly feasible to do more than to indicate some of the best methods of installing a petrol motor plant in a boat, but the author hopes this effort will help his readers to turn out a practical boat.

A Scale Model T.-B. Destroyer.

CHAPTER XI.

THE great advantages of electricity as a motive power for model boats are that it is clean, adaptable, reliable, and always available for immediate use, provided the accumulator has been properly charged. Fig. 192 shows the general arrangement of a typical electrically driven boat, and although the type of vessel may vary, the system remains the same, and the general description here given applies equally to a small model electric motor boat as to a large liner or battleship, as Fig. 212. Of course, it is practically impossible

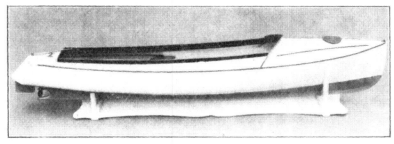

Fig. 192. A Typical Electric Boat.

to get a very high speed with an electric plant, as the size of the accumulators is naturally restricted, and it is not feasible to obtain a sufficiently heavy discharge from them for a real fast racing boat. First, as regards the supply of motive power, this is invariably obtained from an appliance known as an accumulator, which put briefly may be termed a contrivance for storing up electricity, much in the same way as a sponge holds or stores up water. An accumulator consists of a box or casing, usually of celluloid, in which are contained three or more plates of lead, two of which are termed the negative and one the positive plate. These are coated with a special preparation known as paste, and after they have been " formed " they are ready for charging. Accumulators are usually supplied uncharged, and before they can be charged they must be filled with a solution of dilute sulphuric acid, of the precise quantity and density as indicated in the charging instructions, which are generally printed on the accumulators. The process of charging an accumulator consists of passing a continuous current of electricity through the plates of the cell. The action of such current causes a chemical change in the constitution of the plates, and after the current has been flowing for a length of hours, the change in plates becomes complete, and they can then generate a current. This may be observed by noticing that the plates " gas " very freely, that is to say, small bubbles appear on the surface of plate and rise to the top of the acid, the appearance being something like that of aerated water; and further, the positive

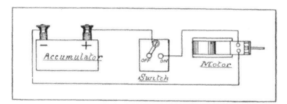

Fig. 193.   Wiring a Simple Electric Boat.

plates have turned a very dark brown colour, whereas before charging they were a much lighter brown. Most electric accumulators consist of a number of cells, each of which has a capacity of 2-volts : thus if four volts are required, it will be necessary to have two 2-volt cells, putting the same in series ; for 6 volts three cells will be necessary, and so on in proportion. In practice, these accumulators are supplied in one case containing the required number of cells, so that it is merely necessary to connect up two terminals. The method of wiring up the ordinary boat motor with a switch and 4-volt accumulator is indicated in Fig. 193, one wire being taken from the positive terminal (this is always marked with a +, or painted bright red) to one of the terminals on the motor, and leading one wire from the negative terminal (this is always marked by a —, and painted black) to one terminal of the

P

switch, another wire being carried from the opposite terminal of
the switch to the remaining terminal on the motor, this being
clearly indicated in the illustration. If a reversible motor is used
the wiring becomes somewhat more complicated, but with the aid
of Fig. 194 it should be easily possible to trace out the "leads," as
the wires leading from battery to motor are called. In boats with
a sufficiently large carrying capacity two separate accumulators

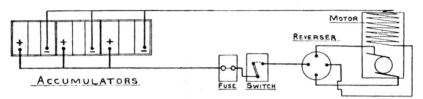

Fig. 194.   Wiring an Electric Reversing Motor.

may be used, in which case a " two-way switch " could be inserted,
enabling one accumulator to be used, and after that becomes
exhausted, switching over to the other. The method of wiring
such a combination is indicated in Fig. 195. It is beyond the scope
of this book to describe in detail the construction of accumulators
or an electric motor, but it may make matters clearer and tend to
a more comprehensive understanding of an electric model boat if

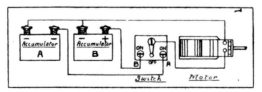

Fig. 195.   Wiring of Two Separate Accumulators with
Change-over Switch.

a brief description of its working is given. The amount of current
that a cell can store or accumulate depends upon the area of its
plates ; the larger the plate in area or number, the greater the
capacity of the cell, and, in addition, the rate of discharge, or the
amount of current that can safely be given off, is increased by
increasing the size of the plates.

The usual sizes of accumulators are given here for reference.

TABLE No. 14.

SIZES OF STOCK ACCUMULATORS.

No.	Volts.	Amp. hrs.	Height.	Length.	Width.	Weight.	
			ins.	ins.	ins.	lbs.	ozs.
1	4	4	$2\frac{3}{4}$	$3\frac{1}{8}$	$1\frac{3}{4}$	1	4
2	4	10	$3\frac{1}{2}$	4	2	2	12
3	6	10	$3\frac{1}{2}$	4	3	4	0
4	4	30	$6\frac{1}{4}$	$3\frac{9}{16}$	$2\frac{5}{8}$	6	8
5	6	30	$6\frac{1}{4}$	$5\frac{9}{16}$	$2\frac{5}{8}$	9	12
6	4	40	$6\frac{1}{4}$	$3\frac{13}{16}$	$4\frac{3}{16}$	11	4
7	6	40 .	$6\frac{1}{4}$	$5\frac{1}{16}$	$4\frac{3}{16}$	16	12

To determine the size of an accumulator to drive a given motor for a given time it is necessary to know the voltage of the motor and its current consumption in "amps.," the voltage indicating the pressure of the current, and the amps. (amperes) the quantity. Thus, for example, the small size " Nautilus " motor requires at 4 volts, 1.5 ($1\frac{1}{2}$) amps. to give full power ; consequently, the No. 2 accumulator, of 4 volts, with a *total* capacity of " 10 ampere hours," that is, one ampere of current can be constantly discharged for ten hours from the plates before the battery is exhausted, but during this time the voltage will drop, until at the end of ten hours there would be no " volts," consequently we must reckon the " amps." only so long as we have sufficient " volts " for the purpose, and it is usual to divide the " amp. hour " capacity by two to obtain the working range of the cell. Thus, at 4 volts (approximately) the No. 2 cell will deliver 1 amp. for five hours, or $1\frac{1}{2}$ amps. for about three hours. It is not practical to discharge this cell at a higher rate than $1\frac{1}{2}$ amps. or the plates might be buckled, due to the violence of the reaction caused by the heavy discharge.

An electric motor consists essentially of four parts, the field magnet, marked FM in Fig. 196, and the armature A, the commutator C, and the brushes B. The principle upon which an electric motor operates presupposes that if a current of electricity be passed round a coil of insulated wire it will induce a magnetic flux, or flow of magnetism, in the core. This magnetic flux is induced in the core of an electric motor by what is known as the field windings

(shaded Fig. 196), which are wound around the core D, this usually
being in the form of the letter U, the induced magnetic flux
flowing through the air from one pole to the other being known as
the " magnetic field." A similar process takes place in the armature, which consists of an iron core wound with a coil, or coils,
of wire ; the ends of the wires being connected to the separate
segments of the commutator, the two brushes serving to lead a
current of electricity from the accumulator, *via* the segments of
the commutator, to the armature coils, where they induce another
magnetic flux of opposite polarity to that of the field, and by the
well-known law of attraction and repulsion cause a movement in
the armature.   The effect of the three or more poles is to constantly
present a changing magnetic flux in the poles of the armature to

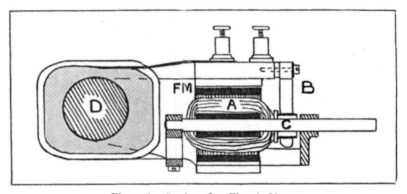

Fig. 196.   Section of an Electric Motor.

the constant magnetic field, continuous rotation of the armature
being the result.   The most common causes of a small motor failing
to work properly are :   insufficient current, improper adjustment of
the brushes, the presence of dirt or other foreign matter on the
commutator, a breakage in one of the wires, or a breakdown of the
insulation.   Of these by far the most common is improper adjustment of the brushes.   The brushes are usually of three forms in
the simpler models : they may consist of two flat strips of copper
as indicated in Fig. 197, and it is necessary with this type of brush
to see that they bear firmly but not too heavily on the commutator,
and that the points of contact are exactly opposite each other, as in
the illustration.   Should they not be exactly opposite, sparking

will take place, causing the motor to run with a loss of power, and in extreme cases the motor would stop altogether. A second common type of brush is that known as the spring plunger type, and illustrated in Fig. 198. These are very simple to adjust, and they are reliable in use, the only other general type consisting of small copper or carbon-copper brushes, mounted on the end of short arms con-

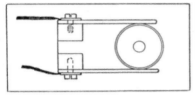
Fig. 197. Flat Brushes of Electric Motor.

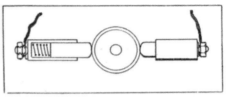
Fig. 198. Spring Plunger Brushes.

trolled by tension springs as illustrated in Fig. 199. The same remarks apply as regards adjusting the pressure and the accuracy of these brushes. In all cases the commutator should be kept clean and free from any dirty matter, but with the *slightest trace* of oil to assist in the reduction of friction. The spindle of the electric motor should also be oiled occasionally with a light oil, but care should be taken not

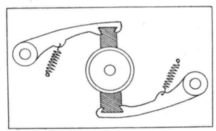
Fig. 199.   Carbon Brush Holders.

to allow the oil to get on to the windings or other electrical parts, as short-circuiting frequently follows such neglect.

There are two principal types of continuous current motors—those known as the shunt wound, and those known as series wound. The difference will be appreciated by studying Figs. 200 and 201. Fig. 200 shows the wiring of a series wound motor, from which will be observed that the current flows first into the field windings and from thence through the brush gear around the armature windings and from thence back *via* the brush to the accumulator. With the

shunt wound motor, Fig. 201, it will be noticed that the current
flows round the field windings and round the armature windings
at the same time, the current from both sets of windings return-
ing to the negative pole of the accumulator.   The chief practical
difference between the two types of winding is that the series motor
will run to a very high speed on light loads, while the shunt winding
will maintain a slower, more regular speed under a heavier and varying load.   To reverse an electric motor it is necessary to reverse

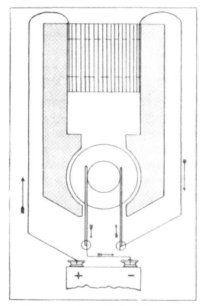 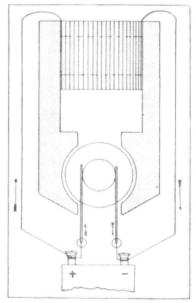

Fig. 200.   Series Motor Winding.   Fig. 201.   Shunt Motor Winding.

the polarity of either the armature or the field, and for this purpose
a four-way reversing switch, as shown in Fig. 202, is necessary.
Reversing apparatus is not usually fitted to model boats; the method
of connection will, however, be quite clear with the aid of the
diagram.

There are now upon the market many very good electric boat
motors, designed expressly for model boat work.

A reliable small motor for boats of 24 in. to 30 in. in length is
shown in Fig. 203.   This measures $3\frac{1}{8}$ in. long, $1\frac{3}{8}$ in. wide, and is
supplied complete with switch, thus making a very neat unit.

A type of permanent magnet motor with reduction gear rendering it very suitable for paddle boats is the " Tracto." This is shown in Fig. 204, complete with reduction gear; when fitted up with flexible connections, it is quite suitable for a paddle boat. The use of the flexible connections allows the motor to be withdrawn without disturbing the paddles.

A well-known and very high-class motor suited to boat propulsion is shown in Fig. 205. This is the ''Fulmen,'' made by A. H. Avery of Tunbridge Wells. The popularity of the type is evident from the number in use, and for convenience in planning a model boat the following particulars are given of these " Fulmen " traction motors.

No.	Suitable for Models Weighing	Dimensions.	Weight.	Current Required.
T 1	10 lb.	$4\frac{1}{2} \times 2 \times 2$ in.	1 lb.	4 v. $1\frac{1}{2}$ a.
T 2	20 ,,	$6\frac{1}{2} \times 2\frac{3}{4} \times 2\frac{3}{4}$,,	$2\frac{1}{2}$,,	6 v. $2\frac{1}{2}$ a.
T 3	56 ,,	$8 \times 4\frac{1}{4} \times 4\frac{1}{4}$,,	10 ,,	8 v. 4 a.

The correct size of electric motor for a given size and type of model boat can readily be ascertained when the resistance of the boat is known. Calculating as described in Chapter IV on the resistance of hulls, the designer can obtain the approximate resistance to be overcome in " Watts." All that remains, therefore, is to select a motor that will give an effective power of the required number of "Watts."

For example, a liner model may require about 36 to 40 "Watts" to drive it at a speed of 4 m.p.h. The motor probably would be wired for 8 volts, the amperage would then be 5, as volts multiplied by amps. gives " Watts." Accumulators of sufficient capacity to safely stand a constant discharge of 5 amps. must therefore be selected, the length of run depending entirely upon the efficiency of the motor and the size of the accumulators.

The controlling apparatus for an electric boat is very simple. The most important is a switch for regulating the speed and reversing the direction of motion. A simple switch of superior construction, as shown in Fig. 206, consists of an ordinary two-terminal " stop and start " switch, but is of superior construction to the usual type, being made of fibre and brass, and represents a small " booby hatch."

The reversing switch Fig. 207 is more elaborate but is neat and compact, and can be fitted in almost any position on the model boat, as the handle only need project above the deck.   The switch is made in the best possible manner, and is perfect in its action.   A regulating switch or controller is not often used in model boat work,

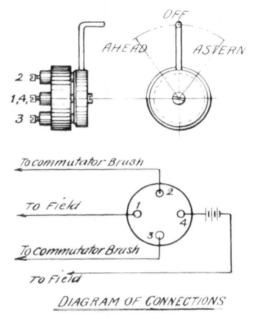

Fig. 202.   Reversing Gear for Electric Motor.

but when used consists of a coil of specially prepared wire having a high resistance, *i.e.* it presents a resistance to the flow of the electric current, consequently the amount that passes can be regulated by inserting more or less of the resistance material, and the motor then runs faster or slower in like proportions.   A simple and cheap pattern is shown in the illustration, Fig. 208.

A " fuse box," Fig. 209, using a soft thin lead wire, protects the accumulators from accidental short-circuiting, as the heat generated by a " short " melts the lead wire and so breaks the current.

In handling a model electric boat of any description the chief points to bear in mind are: first, to keep the accumulators properly charged ; secondly, see that the terminals and accumulator are kept free of any acid which may have splashed out, and also free from dust and moisture. All the connections should be made with copper wire, the ends of which should be clean and bright. All connecting screws and terminal nuts should be screwed down firmly to make a good metallic contact, taking care to see that the switches, stopping and starting levers, etc., work perfectly and smoothly, and that contact is made at all necessary points. The propeller shaft

Fig. 203. Typical Small Electric Boat Motor.

and gears, if any are used, should be kept well oiled and working freely, and it is a good plan when a boat has been taken from the water, before putting it away, to wipe the exterior dry with a cloth, putting a few drops of oil on the moving parts, and disconnecting the wires from the accumulator, this latter precaution frequently saving the same from being accidentally exhausted. Care should also be taken to see that the boat is kept upright, as the acid otherwise may be spilled, causing considerable damage to the interior of the boat. The principal points to be borne in mind when purchasing a model electric boat are that the accumulators must be large enough

to drive the boat at its full speed.   It is no use purchasing a large
electric motor, and expecting a small accumulator to drive it, either
at its full speed, or for any length of time, and to make this matter
clearer the analogy of the sponge should be borne in mind, as no

Fig. 204.   The " Tracto " Boat Motor.

matter how efficient the accumulator may be, it cannot hold more
than a certain amount of electricity, the quantity being technically
known as amperes, the voltage indicating the pressure, and, in the
case of accumulators, being a direct indication of the number of

Fig. 205.   The "Fulmen"
Boat Motor.

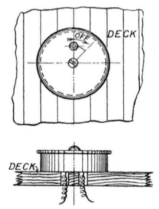

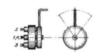

Fig. 207.   Reversing Switch
for Model Boat.

Fig. 206.   Booby Hatch Switch.

cells, and for general purposes it may be assumed that 8 sq. in.
of positive plate will give a constant discharge of one ampere.
Electric motors are wired to take a definite number of amps. at a

known pressure, and this should be ascertained when purchasing. The foregoing remarks apply to electric boats in general.

One of the simplest models on the market is the " Defender," and these boats are provided with a 4-volt accumulator and a small motor. A small switch is provided, and the connections are the same

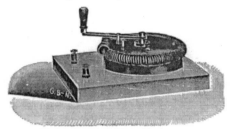

Fig. 208. Coil Type Resistance.

as those indicated in Fig. 193. Torpedo boat destroyers are fitted with a larger motor of somewhat different construction, the current being switched off and on by means of a special switch, constructed to represent a model torpedo tube, shown in Fig. 210, but with this the same instructions should be followed. All of these could be fitted

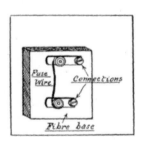

Fig. 209. Simple Fuse Box.

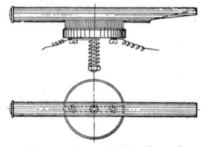

Fig. 210. Torpedo Tube Reversing Switch.

with twin screws, in which case the gears should have a small quantity of vaseline applied to the cogs. This will be found to improve their quietness and ease of running very considerably. To recapitulate, take care to see first that the accumulator is fully charged. See that all connections are cleaned and well made, that the brushes and commutator on the motor are cleaned and properly adjusted, all moving parts sufficiently oiled and working smoothly

and freely. In the foregoing remarks mention has been made of
" short-circuiting." This may be described as directly connecting
or closing a circuit, the simplest illustration of direct short circuiting being that of connecting the two terminals on the accumulator
together ; this would have the effect of tending to immediately
discharge the accumulator, the result of which would be either to
cause it to catch fire, or to buckle the plates, causing the paste to
drop out of the cells or pores in the same, and to entirely ruin the
accumulator. For this reason the two terminals must never be
connected directly together, and for the same reason no two
terminals on the motor or any other part of the circuit should be
directly connected, when they are of opposite polarity, or on the
flow and return circuits. Should the insulation of any of the wires
become chafed or destroyed for any reason, there will be nothing to
prevent the current flowing from the wire to the adjacent metal
work, and if there should be any connection with the return circuit,
a " short " will be the result. Short circuits, however, need not be
anticipated if the model as supplied is not meddled with, and every
precaution taken to properly insulate and separate all parts liable
to short circuiting , so that with reasonable care no danger may be
anticipated, but a safety fuse should be inserted in the circuit as a
safeguard.

Some examples of electrically driven models are given here, and
serve to show that very elaborate models can be made, perfectly
equipped with deck fittings, thus making a very handsome and pleasing boat. Fig. 211 is a reproduction of a fine model, the battleship
H.M.S. *Lord Nelson*, a British warship completed in 1906, and
marking the transition stage between the earlier *King Edward*
and the later *Dreadnought*. It is built to a scale of $\frac{1}{10}$ in.
to the foot, and is 3 ft. 7 in. long, 8 in. beam, $4\frac{3}{4}$ in. deep, and is
fitted with forty-two accurate scale model guns, eight searchlights,
lifeboats, admiral's stern walk, and is arranged to be driven
electrically with twin screws. The " Lowko " motor is utilized,
and drives through a gear box to synchronize the revolutions of the
shafts. The motive power is supplied from a 6-volt accumulator,
having a charge sufficient for a run of an hour's duration. The
model is painted in standard grey, and the illustration gives a good
idea of this fine boat, built by Bassett-Lowke from designs by the
author. This model, the property of Mr. R. L. Robinson, has

Fig. 211.   Electrically-driven Model Battleship, **H.M.S.** *Lord Nelson.*

Fig. 212.   Scale Model Battle Cruiser *Queen Mary.*

gained a premier award in open competition for spectacular models.

Another fine vessel is the 6 ft. 6 in. battle cruiser *Queen Mary*, shown in Fig. 212. This is completely fitted and is the property of Lord Howard de Walden.

Another great advantage of an electric boat is the ease with which the navigation lights, searchlights, and so forth, can be illuminated, by means of a " Pea " lamp usually supplied with current from a small lighting battery or accumulator of 4-volt capacity. The "pea" lamp P, as purchased, has two wires projecting, and these should not be broken ; the lamp is to be mounted on a small block of fibre or hard wood F, with two holes drilled through same for the wires aforementioned, the lamp being secured with a touch of sealing wax, rubber solution or white lead, as shown in Fig. 213 at R. The head lights and side lights are readily constructed from short circular brass tubes with lamp and fibre base pressed into the bottom, and a wood or metal cap fitted on top, as shown in Fig. 214.

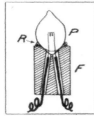

Fig. 213. Fitting a Pea Lamp to Socket.

Searchlights are readily made, for simple working models, by turning the body of lamp as at B Fig. 215, from hard wood, painting the interior white to form a reflector which should be parabolic in form. The lamp is fitted by drilling a hole through the body, at right angles, as shown at C, and cementing or simply pressing the fibre block into place. The front is glazed with a circular glass disc, the base turned to shape from hard wood, and the lamp supported by a simple U-shaped bracket, as indicated at A.

Of course, more elaborate fittings can be made if desired, and those readers intending to do so are referred to Chapter XII, dealing with deck fittings in detail.

The wireless control of a model boat is quite possible, and has been successfully carried out by Capt. Colston, to mention only one experimenter, but the number and delicacy of the instruments, and their comparative costliness, together with their weight, render the extensive adoption of this fascinating control beyond the scope of ordinary power boat users, although it is most interesting as a study.

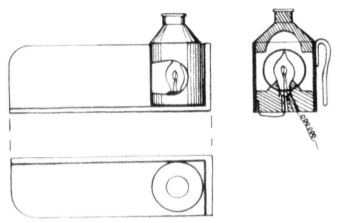

Fig. 214. Simple Electric Head and Side Lights.

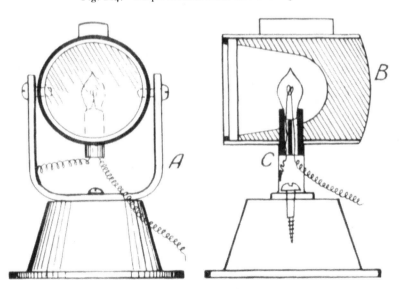

Fig. 215. Simple Model Electric Searchlight.

CHAPTER XII.

DECK FITTINGS.

THE term " deck fittings " embraces a very wide range of detail model work, and includes such diverse items as the big guns of a model warship to such a simple fitting as a plain tubular funnel for a little steam tug ; but this subject may broadly be divided into two classes :—

A. Exhibition fittings, such as are exclusively used on glass, case models.

B. Practical fittings for use on boats that are intended for actual use on the waters of a pond.

The first category of exhibition fittings require the most delicate construction, the provision of fully detailed drawings, and an absolutely faithful adherence to scale and proportion. Such fittings are necessarily fragile, and quite unsuited to proper working models. They are, moreover, subject to the most discriminating criticism when displayed in a public museum, and it is not to be wondered at that the construction of such fittings requires exceptional ability and patience, acquired only by long experience and apprenticeship to a practical firm of ship modellers.

On the other hand, the modelling of practical fittings for a working model is not such a formidable undertaking, as provided the *character* is retained, all merely superfluous detail can with advantage be entirely omitted, the result being a more robust fitting, suited to arduous service under actual working conditions.

A compromise is frequently effected by fitting up an electric working model with the more important deck fittings, such as anchors and winches, accurately modelled, and finished in such a high-class style as to be worthy of exhibition use, although the tiny details are eliminated, as under working conditions they would be entirely ruined.

An excellent example of exhibition model work of the highest class is shown in Fig. 217; this shows the latest British battle cruiser *Queen Mary*. The model is about 14 ft. long, and was built by Messrs. Palmers Shipbuilding Co., and presented to Her Majesty the Queen. Every detail is beautifully carried out, as befits a model intended for such high honour.

Some wonderful examples of ship modelling are to be seen in the Science Museum at South Kensington, while the apparently unending aisles in the Naval Section of the Musée de Louvre, in Paris, provide the spectator with a unique history of shipbuilding from the earliest times. Another magnificent collection of ship models is to be found in the Munich Museum, while at the Congress House at Washington, U.S.A., is, or was, a collection of early and contemporary U.S.A. naval models of the highest class.

Fig. 217. Scale Model H.M.S. *Queen Mary*.

The fine model Australian mail boat *Omrah*, of the Orient Steam Navigation Co., built by the Fairfield Shipbuilding and Engineering Co., Ltd., and on exhibition in the Science Museum, Kensington, is shown in Fig. 218. She is a twin-screw boat 507 ft. in length, and would make a splendid prototype for a practical working model, on which the deck fittings could be numerous and accurate; the massive single funnel, the two masts, and neat arrangements of upper deck being particularly advantageous from a modelling point of view, as the whole central structure could be detachable to give access to the interior mechanism. A typical cargo boat with corrugated hull, illustrated by Fig. 219, gives an excellent idea of the fittings on these boats, but a more imposing model is shown in Fig. 220. This is the

Q

Fig. 218. Exhibition Model SS. *Omrah*.

famous liner *France*, owned by the Compagnie Générale Trans-atlantique. The model is two metres in length and replete with all fittings, modelled in detail and exact to scale. She has quadruple screws, and is a typical example of modern luxury in ocean travel.

Fig 219. Exhibition Model Cargo Boat.

The interesting model—historically—shown in Fig. 221 is an exact scale model of Sir Ernest Shackleton's exploration ship *Nimrod*. This was built with the utmost fidelity and accuracy in every detail ; the author, assisted by a draughtsman, preparing the

Fig. 220. Scale Model Liner *France.*

working drawings from measurements, sketches, and photographs taken from the actual ship on her return to the Thames after her adventurous career.

As it is impracticable to give detailed drawings of all the deck fittings used on the many different types of boat the following alphabetically arranged description of those fittings, which may be termed " standard " for most ship models, has been prepared in the confident anticipation it would be of general utility.　This list is followed by diagrams showing the usual arrangement of the prin-

Fig. 221.　Scale Model Sir Ernest Shackleton's Ship *Nimrod*.

cipal fittings on different types of craft, as although almost every individual boat has a slightly different arrangement of her deck fittings, the diagrams show the general or average disposition of the separate items.

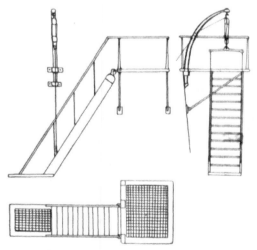

Fig. 222.　Accommodation Ladder.

The following fittings are useful for high-class working or spectacular models, or for ordinary exhibition purposes. No details of the methods of their construction are given beyond mentioning that the principal tools required are a small accurate lathe, hand tools,

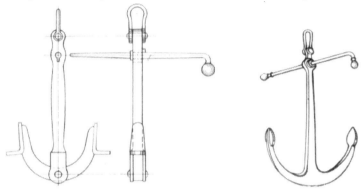

Fig. 223.   Trotman's Anchor.   Fig. 224.   Rogers Anchor.

small blow pipe, and above all care and patience. The metal used to the exclusion of practically any other is brass, while the finish is generally silver or gold plating, bronzing or oxydizing and lacquering in colour, a coat of clear lacquer completing the work.

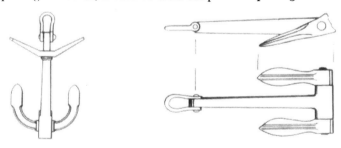

Fig. 225.   Martin's Anchor.   Fig. 226.   Stockless Anchor.

An Accommodation Ladder, Fig. 222, is a hanging staircase or ladder suspended from small davits on the ship's side, provided with wooden treads, or stairs, and a landing platform at top and bottom, this being in the form of a grating. Stanchions and hand-rail complete the work. These ladders are used on large and small vessels

of all types, but they are only lowered when the vessel is in harbour
or at anchor.

The Admiral's Sternwalk, is used exclusively on warships, and is
an erection at the extreme stern of the vessel.   On some modern
British warships, the admiral's walk was arranged between the legs
of the forward tripod mast, and abaft the funnel, but the heat and
dirt in this position speedily dictated a return to the more comfortable

Fig. 227.   Breakwater.

sternwalk.   The illustration in conjunction with the photographs
of finished models, will give a good idea of its arrangement.

Anchors.—As on an actual vessel the provision of suitable anchors
and their attendant gear is an absolute necessity, no model boat with
any pretensions to realism should be without anchors.   Various
types are in use, one of the principal patterns being shown in Fig.
223.   This is known as a " Trotman's " anchor, and is extensively

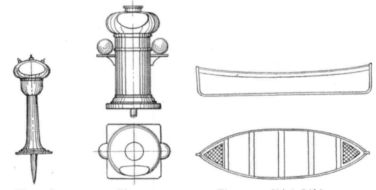

Fig. 228.   Fig. 229.   Fig. 231.   Ship's Lifeboat.
Plain Binnacle.   Thomson's Binnacle.

used on merchant vessels and yachts.   The cross-bar is detachable,
making storage on deck comparatively simple.   Another type of
useful anchor, chiefly used on launches and small steam yachts,
shown in Fig. 224, is known as the " Rogers "; it is neat, simple and
effective in use.

" *Martin's* " *Anchors*, Fig. 225, are largely used in the Navy and
on large ocean-going vessels, but are not suited for stowing in hawse

pipes, owing to the cross-bar or stock. For these reasons many different forms of " Stockless " anchors have been devised. A popular and reliable type is illustrated in Fig. 226, and known as the " Byers." It is in very great use on warships, merchantmen and ocean-going craft of all kinds, especially when, as is now the general custom, it is desired to carry or stow them in the hawse pipes. All the foregoing anchors are usually obtainable in three sizes, suitable for models of different scales, the sizes being as follows :—

STANDARD ANCHOR SIZES.

$\frac{1}{10}$ in. scale. $1\frac{1}{8}$ in. long.

$\frac{3}{16}$,, ,, $1\frac{1}{2}$,, ,,

$\frac{1}{4}$,, ,, $1\frac{3}{4}$,, ,,

Awning Stanchions are of many different types and sizes, but consist essentially of a tall rod or stanchion, arranged to carry the

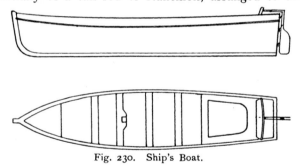

Fig. 230. Ship's Boat.

awning or sun-screen on boats voyaging to tropical parts. The outer or deck side stanchions are plain, and those destined to hold the ridge or centre beam, and fitted in the centre line of the ship, are provided with a forked head, to support the ridge pole.

A Breakwater, Fig. 227, is fitted near bows of boat to prevent water washing over deck.

Binnacles.—These are sometimes known as compass stands, and are of various types. The plain or simple pattern is shown in Fig. 228, and used on small craft, or the improved and patented pattern such as the " Thomson's " Fig. 229. They are essential fittings on the navigating bridge, as without them the proper working of the ship would be impossible.

Boats.—The proper adequate provision of ships' boats on ocean-going vessels has been vividly before the public recently, and the up-to-date model enthusiast should see that his model is not lacking in this respect.

Ships' Boats are many and various in design, but broadly speaking, for model work, are of three types :—

A. Open boats with transom stern as in Fig. 230.

B. Open boats of the " lifeboat " type with both ends pointed as in Fig. 231.

C. Covered boats as used on ocean-going vessels, when at sea. These are of course only the ordinary open ships' boats, but protected by a canvas covering.

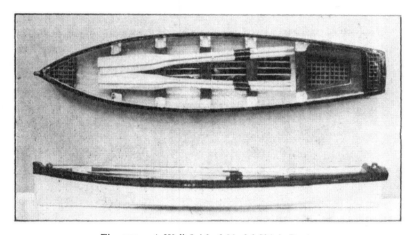

Fig. 232.   A Well-finished Model Ship's Boat.

The open boats may be quite simple in construction, and cut out of a little piece of pine, with two or three cross pieces to represent the thwarts or seats, or may be elaborately finished with knees, treads and oars, as in Fig. 232. The covered boats are very simply made, and are generally left solid. The proportions of the boats, of course, do not vary with any variation of their scale, so that by following the lines given in the drawings a serviceable and accurate boat will be obtained.

Dinghys are, however, generally rather broader and deeper, while gigs are slightly narrower in proportion to the length. The average dimensions of a ship's lifeboat to carry fifty persons is 28 ft. long, 8 ft. 6 in. beam, 3 ft. 6 in. depth, although longer and shorter boats are used occasionally. The dimensions given are those issued in June, 1913, in the Report of the Departmental Committee of Boats and Davits.

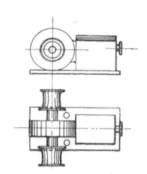

Fig. 234. Boat-lowering Engine.

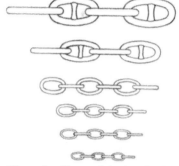

Fig. 238. Types of Anchor Cable.

On warships it is customary to provide a number of steam launches, usually about 40 ft. to 60 ft. in length, and styled variously as steam pinnaces, vedette boats and launches. These are used for harbour service, mine laying, towing boats for a landing expedition, and a hundred and one other purposes. They are fast handy boats, a typical design being given in Fig. 233, of a small model 5 ins. long ($\frac{1}{10}$ in. scale of a 60 ft. boat).

Fig. 233. Typical Steam Pinnace.

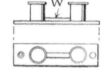

Fig. 235. Vertical Bollards.

Boat-lowering Engine.—This is an electrically driven machine for speedily raising and lowering the boats, generally used in conjunction with the " Welin " and such-like davits. A simple model is shown in Fig. 234, such as largely used on the big liners.

Bollards are used for mooring purposes to make fast the hawsers which hold the vessel to quay side, and for other purposes. They are plentiful on the actual vessel, and as they are cheap, and easily made, can with advantage be freely used in their proper places on the model.   Fig. 235 is the vertical pattern.   Fig. 236 shows a powerful staggered or self-jambling pattern; both are in extensive use on shipboard.

Boom Fittings.—The many derricks and booms used on shipboard necessitate considerable rigging, and to facilitate this some amount

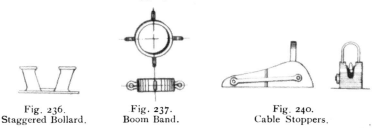

Fig. 236.   Fig. 237.   Fig. 240.
Staggered Bollard.   Boom Band.   Cable Stoppers.

of metal work is required ; that most generally used being a metal band with four eyes, as shown in Fig. 237.   These are supplied as stock fittings in all sizes from $\frac{3}{16}$ in. to 1 in. dia.

Cable.—This is obtainable in a variety of styles, the simplest being the common round link, bent to shape.   An improvement is to solder

Fig. 239.   Cable Gear and Senhouse Slip.

the ends of each link together, while the best and proper anchor cable should be studded, that is, should have a little cross-bar soldered in place across each link.   These various styles are shown in Fig. 238.

Fig. 241   Merchant   Fig. 242.   Fig. 243.
Marine Capstan.   Steam Capstan.   Warship Capstan.

Cable Gear (the general term for the gear used to make fast a cable to a riding bitt or a standing stopper bolt) is shown in Fig. 239, but usually is only fitted to large or elaborate models.

Cable Stoppers or controllers are always used to check a cable when running out, and to serve as a riding bitt on ordinary occasions. It consists, as can be seen by Fig. 240, of a metal body with groove through which the cable passes. A hand lever at the side controls the movement of a trigger, which engages with the links of the chain, and so stops its movements. The hoop over the top prevents the chain jumping.

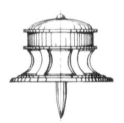

Fig. 244. Cruiser Capstan.

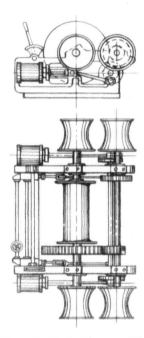

Fig. 245. T.B.D. Capstan.

Fig. 246. Clarke Chapman Winch.

Capstans are used to hoist in an anchor by winding in the cable, or to lift heavy weights, warp a boat alongside a wharf, and so forth. There are many different patterns, some of which are here illustrated. Fig. 241 shows the ordinary merchant service type, as used on small vessels, while on larger craft a steam capstan such as shown in Fig. 242 is extensively used on deck, the capstan engines being stowed beneath the same. The usual heights for such fittings are as follows :—

$\frac{1}{10}$ in. scale, $\frac{1}{2}$ in. high.

$\frac{3}{16}$,, ,, $\frac{5}{8}$,, ,,

$\frac{1}{4}$,, ,, $\frac{13}{16}$,, ,,

The modern warship, with its huge anchors, requires very substantial capstans to deal expeditiously with the loads, and patterns

Fig. 247. Photo of Model Winch (Full Size).

suitable for warships (Fig. 243) or cruisers (Fig. 244) are shown and are thoroughly characteristic models.

The type illustrated by Fig. 245 is largely used on destroyers and such craft, being somewhat lighter than the warship capstan. The type modelled has circular handle on top to control the steam gear, and is a very attractive and accurate addition to a model.

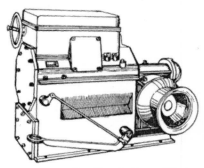

Fig. 248. Electric Cargo Winch.

Fig. 248A. Cargo Hatch.

Cargo Winches are made in a variety of sizes and types, but that depicted in Fig. 246 is the general type and represents standard practice by the well-known marine engineers, Messrs. Clarke, Chapman & Co., and is frequently met with on exhibition ship models. Their purpose is to expeditiously raise cargo, etc., from the holds of a boat, access to the interior of the hold being by means of removable coverings, known as hatches.   Fig. 247 is a photo of such a fitting.

On large modern liners, electric cargo winches, Fig. 248, are very extensively used.

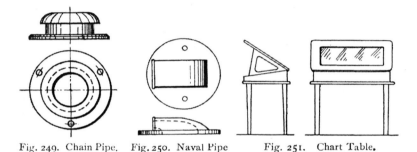

Fig. 249. Chain Pipe.   Fig. 250. Naval Pipe   Fig. 251.   Chart Table.

Cargo Hatches may readily be made of thin mahogany.   Fig. 248A shows the general style, and the usual stock sizes are as follows :—

> Length $2\frac{5}{16}$ in.   Width $1\frac{3}{16}$ in.
>   ,,   3   ,,   ,, $1\frac{3}{4}$,,
>   ,, $3\frac{1}{2}$,,   ,, $2\frac{1}{4}$,,
>   ,,   4   ,,   ,, $2\frac{3}{4}$,,

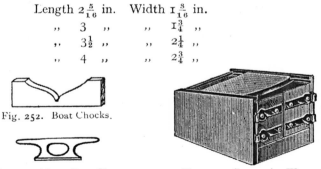

Fig. 252.   Boat Chocks.

Fig. 253. Motor Boat Cleat.   Fig. 254. Companion Way.

Chain Leads and *Chain Pipes* are used to convey the cable from the chain lockers in the forepeak, to the capstan or windlass and thence to the anchors. The general type, as used on merchant vessels, is shown in Fig. 249, and is circular in plan, while on a battleship a more elaborate device known as a naval pipe is used. Fig. 250 shows the arrangement of this fitting, which is characteristic of naval practice.

Chart Table.—A neat fitting on which the chart is placed under cover of a sloping glass or metal shield, the arrangement being as shown in Fig 251.

Chocks are used under boats and other fittings to prevent their moving, or make a secure fixing for them.   Boat chocks are shown

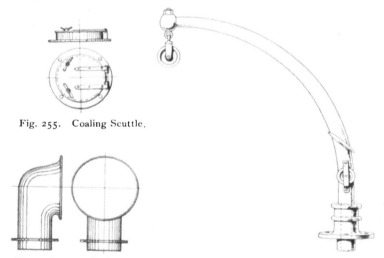

Fig. 255.   Coaling Scuttle.

Fig. 256.   Ventilating Cowl.   Fig. 257.   Anchor Davit.

in Fig. 252, and are in most extensive demand on liner models, with a large number of boats stowed on deck.

Cleats, used to make fast a sheet or running rope, temporarily, are not in extensive use on power models, but are frequently found on motor boats, a standard fitting of this class being illustrated by Fig. 253.

Companions are used to enable persons to have access from one deck to another.   They are usually made in wood, and Fig. 254 gives an excellent idea of their appearance.

Coaling Scuttles, as their name implies, are fitted on deck over the coal bunkers, and are usually circular in plan, Fig. 255 giving details of their arrangement.

Cowls, sometimes called ventilators, are provided to admit air to the interior of a vessel; usually the expression is confined to the smaller sizes of vents. Fig. 256 indicates a practical design which is entirely open and therefore suited for working models. The stock sizes range from ⅜ in. to 1 in. dia. of pipe.

Davits are used chiefly to hoist in ships' boats, but sometimes for dealing with cargo, anchors and the like. Fig. 257 shows one type of single-arm anchor davit, while the more usual and heavier pattern is illustrated in Fig. 257A. Boat davits were at one time exclusively

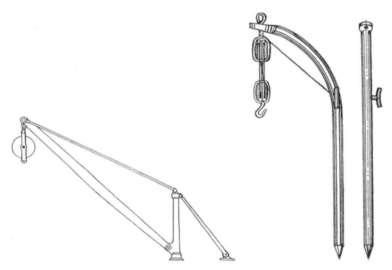

Fig. 257A. Anchor Davit.   Fig. 258. Simple Davit.

of the regulation plain bent pattern as in Fig. 258, which is made of brass wire nicely bent to shape, and drilled at the top to take the block.

A superior type, shown in Fig. 259, has proper tapering arm with ball and chain plate at top and is used extensively on merchant and naval vessels. These are fitted to the boat by various means. On destroyers the usual fitting is a webbed vertical support, as in Fig. 259. For davits on the ship's side, a deck or rail plate Fig. 260, and socket as in Fig. 261 are used, the arrangement being made clear by Fig. 262, showing an ordinary ship's davit in position on the boat's side,

complete with blocks and falls and having flat deck plates B and
footstep brackets C. These can be constructed from thin sheet metal
and the davits D turned from $\frac{1}{8}$ in. dia. brass rod. The block and
tackle at A can be purchased for a few pence ready made, although
if desired the blocks can be cut to shape from small pieces of boxwood. In the case of boats stowed on an upper deck, the rail plate

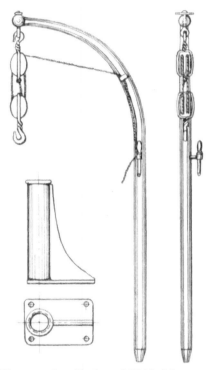

Fig. 259. Boat Davits and Webbed Supports.
As used on **T.B.** Destroyers.

is used on upper deck, and a footstep or base plate on the lower
deck. When rail or deck plates are not convenient, side sockets,
Fig. 263, can be used in their stead.

In more recent times the " Welin" type davits, Fig. 264, have been
extensively used by our great steamship companies. They are
splendid characteristic models, and most handsome and realistic
fitments to any model.

Deck Tubs and *Buckets* are generally provided as an aid in extinguishing fire, and also for use when swabbing the decks. They are simply made of hard wood, Fig. 265 showing a typical bucket rack.

Deck Lights are thick clear-glass fittings to admit light to a cabin

Fig. 260.
Davit Rail Plate.

Fig. 261.
Davit Socket.

Fig. 263.
Davit Side Socket.

or beneath the deck, and are generally small, therefore seldom shown on small or working models.

Derricks for handling cargo, boats, etc., are usually confined to cargo boats and warships. They are readily constructed of thin

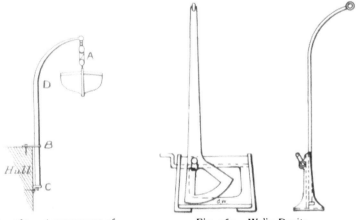

Fig. 262. Arrangement of
Boat Davits.

Fig. 264. Welin Davits.

tube, and are effective fittings, the general arrangement being shown in the photos of typical exhibition models.

An Ensign Staff is used to carry the ensign of a warship, and such fitting is clearly shown in Fig. 266.

Eye Plates are used for attaching the standing end of rigging and

R

consist of a flat base plate with eye bolt securely attached thereto, as in Fig. 267.

Fairleads are devices to guide the hawsers of a vessel without chafing over the ship's side. The mercantile patterns are generally plain, as in Fig. 268, while the pattern Fig. 269 is frequently used on large vessels, the central vertical pulley reducing the friction on

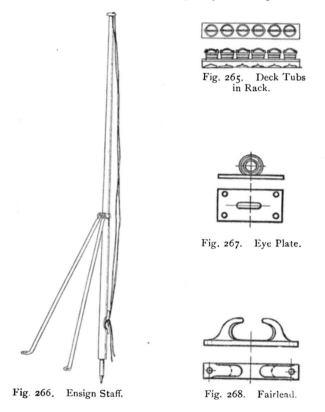

Fig. 265. Deck Tubs in Rack.

Fig. 267. Eye Plate.

Fig. 266. Ensign Staff.

Fig. 268. Fairlead.

the hawsers considerably. Naval type fairleads, Fig. 270, are of a different pattern, the very latest type of fairlead as used in H.M. Navy, and having a dropped flange fitting over the hull sides, being illustrated. These make a fine characteristic fitment to a model warship.

Flagmasts require to be very neatly made or they look clumsy, but are important fittings. On motor boats a mast socket, Fig. 271, is more convenient, a short stout mast with flag being screwed in place when required.

Funnels.—Probably no other deck fitting requires such careful

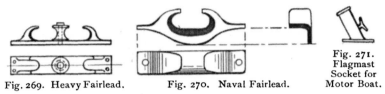

Fig. 269. Heavy Fairlead.   Fig. 270. Naval Fairlead.

Fig. 271.
Flagmast
Socket for
Motor Boat.

design, as a funnel practically makes or mars the appearance of a boat. Their types are legion, and it is impossible to illustrate them all. Fig. 272 makes the characteristics of the general type clear ; thus A is a plain funnel, simply made of tube with a bead at the top and flat rectangular base plate at the bottom, a type of funnel only used on quite small craft.

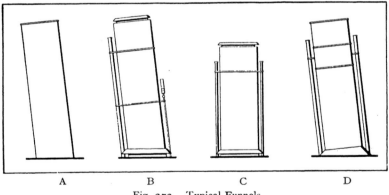

A   B   C   D

Fig. 272. Typical Funnels.

B is an elaborate yacht's funnel with liner, bonnet and capping at base, also steam pipe, and dummy whistle—such a type being frequently found on light cruisers and such craft.

The warship funnel is generally vertical, and C is a good example of a plain funnel of this type.

Liners' funnels are always staggered or lean backwards, and D shows such a funnel, with its banding and dummy pipes, also the usual conical base.

All the above patterns are to be found both circular and oval-shaped in plan. On all types funnel stays are necessary to keep them in position, and prevent their destruction when the boat rolls.

Fig. 273. Gin Block.   Fig. 274. Fiddley Gratings.

The great steamship companies always paint their funnels in a distinctive manner, and Table 15 gives particulars of some of the m ore important, as want of space precludes giving others.

TABLE No. 15.

PARTICULARS OF COLOURING ON LINER FUNNELS.

A   Top of funnel.
B C D   Banding.
E   Body of funnel.
F   Base.

The colour mentioned under these letters indicates the colour at that approximate place on the funnel.

Name.	A	B	C	D	E	F
Cunard . . .	black	black	black	black	red	black
White Star . .	black	—	—	—	buff	black
Hamburg-America .		all buff			—	—
Royal Line . .	blue	—	—	—	buff	black
P. & O. . .		all black			—	—
Compagnie Générale .	black	—	—	—	red	black
Nord. D. Lloyd .		all buff			—	—
Castle Line . .	black	black	black	—	red	black
American Line .	black	white	black	—	black	black

Gin Blocks, Fig. 273, are little metal pulleys with sheave, used in handling cargo, etc. Their construction is plain from the illustra· tion.

Gratings of various kinds are in considerable use on shipboard. Fiddley gratings, Fig. 274, are provided over the boiler rooms to admit light and air, and are made with a metal frame and round wire cross-bars.   Boat gratings are of wood, and for simple working models thin perforated cardboard makes a good substitute for the more expensive wooden kinds.

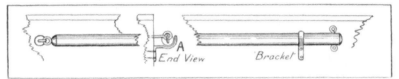

Fig. 275.   Guest Warp Boom.

Guest Warp Booms, Fig. 275, are fitted on the side of a warship (or other vessel), and when extended horizontally allow visiting boats to make fast, and their crew to come aboard by swarming up a rope ladder.

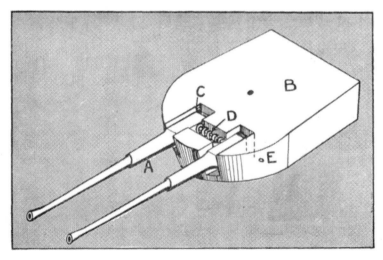

Fig. 276.   Model Big Guns in Gunhouse.

Guns.—The principal feature on a model warship. The big guns of 15 in., 13.5 and 12 in. bore can best be modelled as indicated in Fig. 276. On the actual vessel these are, of course, constructed through- out of the finest tempered steel, but on a model the gun barrels

MODEL POWER BOATS

A may be made from boxwood. They should be square where fitted into the gunhouse B, but turned to correct shape outside. The gunhouse B may be cut from solid block of best yellow pine, two slots C being cut to accommodate the square ends of the gun barrels, a small coil spring D being recessed into place as shown, while a pin E, driven through from the outside and passing right through the two gun barrels, forms the pivot on which the guns elevate and depress, while the spring D forces the guns against the sides of the gunhouse, and holds them in any desired position. It should be finished throughout in naval grey, and be ultimately screwed in place from underneath the deck, the screw being inserted into the central hole at the bottom of the gunhouse, which is shown upside down in the illustration for the sake of clearness.

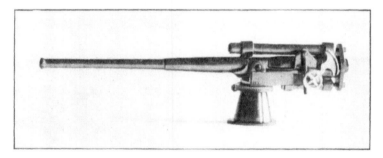

Fig. 277.   Scale Model 6 in. Gun.

Fig. 277 is an illustration of the usual 4 in. or 6 in. quick-firer without shield, and the dimensions for such models are

$\frac{1}{16}$ in. scale, $2\frac{1}{16}$ in. long.

$\frac{3}{16}$,, ,, $3\frac{1}{4}$,, ,,

$\frac{1}{4}$,, ,, $4\frac{3}{8}$,, ,,

These guns are nowadays generally employed behind a shield in the battery, and may therefore be readily modelled as in Fig. 278. The gun shields can be turned from hardwood, and are circular in section, as will be seen from the figure. The gun barrel should be turned from aluminium or brass rod, and screwed into the blocks forming the gun shields. A neat finish is obtained if they are made of brass by bronzing them a steel grey, and finishing with a coat of clear lacquer. The guns are very simply fixed by means of round headed screws, passed through a hole drilled in an angle piece fixed just inside the gunport as shown.

A Gyro Compass is an up-to-date little fitting for the navigating bridge, and Fig. 279 gives an idea of its general appearance.

Hatchways of the water-tight variety, as used on destroyers and

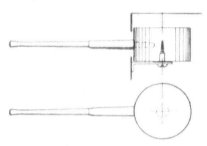

Fig. 278. Simple 6 in. Gun in Housing.   Fig. 279. Gyro Compass.

boats subjected to heavy seas, are illustrated, and are of the rectangular variety, Fig. 280, or circular.

Hawse Pipes are provided at the bows of a vessel, to readily permit the hawsers or cable for the anchor to be handled. Fig. 281 indicates these in position on a boat. The method of fixing them is to mark off their upper position on the deck, and to mark the

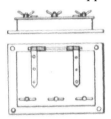
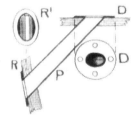

Fig. 280. Watertight Hatchway.   Fig. 281. Fitting Hawse Pipes.

lower position on the outside of the ship's hull, correctly drilling holes from these points in a straight line until they meet in the centre, afterwards enlarging the hole until a tube can be passed straight through, when the projecting ends should be filed almost flush with the deck and the ship's side. A half-round ring R must be fitted on the outside, the front view being shown at R¹. It will be found that the end of the tube will appear to be oval in shape, due to the angle at which the hawse pipe P emerges from the hole. On the deck a circular ring D is fitted, which must be soldered in place when the hawse pipe is finally fitted.

Fig. 282 is the usual merchant type of hawse pipe mouth, while the naval pattern is depicted by Fig. 283. On large vessels the

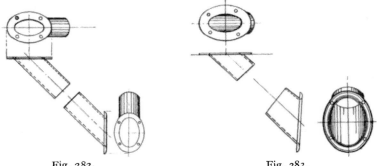

Fig. 282.
Merchant Service Hawse Pipe.

Fig. 283.
Naval Hawse Pipe.

mouth is sometimes made of a shape to suit the head of the anchor. These are difficult to make and fit, and a needless refinement for an ordinary working model.

Headlights, Fig. 284, are used at the mast head to show a white light forwards.

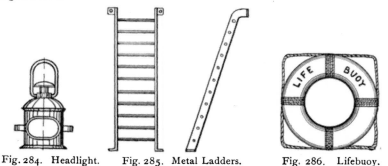

Fig. 284. Headlight.   Fig. 285. Metal Ladders.   Fig. 286. Lifebuoy.

Jackstaff.—The small flag pole at the bows, a neat little fitting.

Ladders are simple and well known, two types being in general use on shipboard :—

    A. Those with round rungs and made entirely in metal, Fig. 285.

B. Ladders having flat treads and made of wood, a more elaborate ladder becoming a staircase, such as is provided on a liner for passenger service.

Lifebuoys, Fig. 286, are for use in saving life, when a person falls overboard.

Masts.—Next in spectacular importance to the funnels is the mast work on a model boat. If the mast and rigging is clumsy

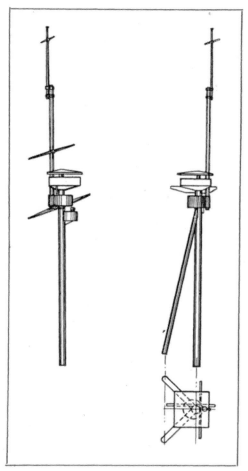

Fig. 287.   Cruiser Mast.   Fig. 288.   Tripod Mast.

and heavy, the whole effect of the boat is spoiled.   The great
aim should be to employ neat, light, graceful mast work.   The
mast itself may be made of hardwood, and rigged as indicated.
The various fittings are all separately described in this chapter.   On

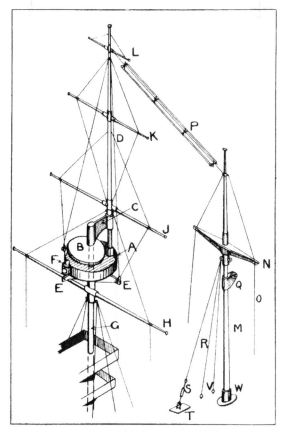

Fig. 289.   Arrangement of Modern Warship Mast and Rigging.

modern warships the masts are of the plain pole variety as shown in
the photo, reproduced on page 233, of the *Queen Mary*, and the
King George V ; while Fig. 287 shows the usual light cruiser masts.
The tripod mast, Fig. 288, is still in very extensive use **on our**
contemporary warships, and is readily made with thin light brass

tubing. The average heights of the warship masts are as follows :—

Tripod Main Masts.			Cruiser type. Pole Masts.		
Height 20 in.	26 in.	36 in.	Height 19 in.	25 in.	35 in.
Scale $\frac{1}{10}$,,	$\frac{3}{16}$,,	$\frac{1}{4}$,,	Scale $\frac{1}{10}$,,	$\frac{3}{16}$,,	$\frac{1}{4}$,,

The general arrangement of the masts on a modern warship is shown in Fig. 289.

The fore-mast is a massive single-pole structure with a circular fire-control station A, with an oval shaped roof B. Above this is a

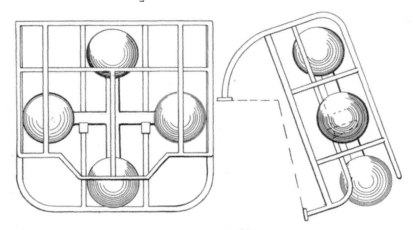

Fig. 290. Night Lifebuoy.

strong bracket C, which supports the top-mast D, while a star piece with four arms E forms the foundation for the fire-control station. The whole of this part can readily be constructed from thin sheet metal. Fig. 289 indicates the general arrangement and proportion of all the parts. The mast-head lamp F is shown in its correct position, and may be turned up from the solid rod, unless it is desired to electrify same, a description of it being given in Chapter XI dealing with the electrification of a model. The mast itself is $\frac{7}{16}$ in. dia. for a $\frac{1}{10}$ in. scale model, and may be constructed from hardwood, and, as will be seen at G, passes through the forward bridges as already mentioned in a previous chapter.

The crossyards H, J, K and I can likewise be fashioned from hard-wood, and may be attached to the mast by cross-pieces of thin brass tube about ⅜ in. long. These can be silver soldered together at right angles, and provide a strong and characteristic fitting for the yards.

The rigging is shown simplified in Fig. 289, and is easily carried

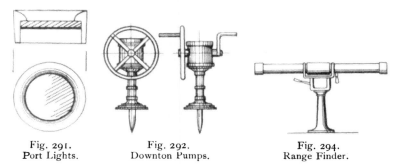

Fig. 291.
Port Lights.

Fig. 292.
Downton Pumps.

Fig. 294.
Range Finder.

out with the special rigging cord now obtainable from most model shops ; the Corinthian brand is specially recommended. The main-mast M is much shorter, and has one metal crossyard N, which serves as a spreader for the top-mast stays O, the principal function of this mast being to support the wireless antennae P, which may be constructed in a most realistic manner from fine silver-plated

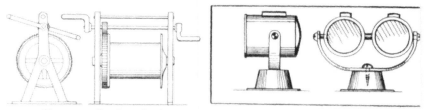

Fig. 293.　Purchase Reel.

Fig. 295.　Duplex Searchlight Projectors.

wire ; the other function of the main-mast being to provide a purchase for the main boom, which is used to lift the various boats, and to lower them overboard. The block attached to the main-mast for this purpose is shown at Q, and may likewise be constructed from thin sheet metal, the sheave being turned from round brass rod. The shrouds R may either be fitted with a small wire strainer S and shackle plate T fitted to the deck, or may be permanently

attached to a small screw-eye screwed into the deck as shown at
V. This is quite satisfactory and neat for a working model. As
the deck supporting the main-mast is frequently removable, a
circular metal plate with short length of brass tube silver soldered
to it must be made to form a support for the mast as shown at W.

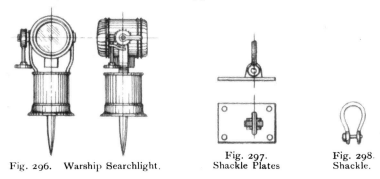

Fig. 296. Warship Searchlight.   Fig. 297.   Fig. 298.
                                         Shackle Plates   Shackle.

The masts should be finished in a lighter shade of grey than the
rest of the boat.

The Night Lifebuoy, Fig. 290, is a prominent fitting on modern
vessels. It consists of a framework of light metal with four copper
balls to ensure buoyancy. An automatic flare is attached, which
on coming into contact with water shows a light, thus enabling the

Fig. 300. Sidelight.   Fig. 301 Skylight.

man overboard to see the whereabouts of the lifebuoy. Its construction for model purposes is quite simple.

Ports are glazed openings in a ship's side to admit light and air.
Fig. 291 is a section of the usual model fitting.

On the outer side of the hull a little fitting known as a Rigol,

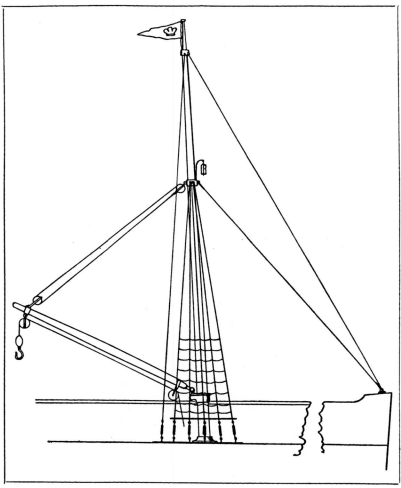

Fig. 299. Simplified Rigging for Fore-mast.

is attached to prevent drainage water from running down the hull sides and entering the port when open.

Flanged Lights are a modified form of port, but generally used in cabin sides.

Pumps.—Downton Pumps, Fig. 292, are universally used to provide against leakage of the hull, and more frequently to supply

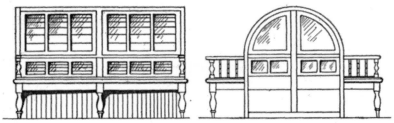

Fig. 302. Skylight with Seats,

water for deck-washing, etc. The construction is simple and readily effected from brass rod and wire.

Purchase Reel, Fig. 293.—A form of small crab. or winch, used for soundings and light haulage and lifting purposes on deck.

A Range Finder, Fig. 294, is fitted on modern warships to ascertain

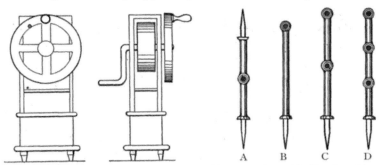

Fig 303. Sounding Machine.   Fig. 304. Typical Railing Stanchions.

the range of one vessel from another, to aid the gunlayers in sighting their weapon.

Searchlights are now in considerable demand on liners, for assistance in berthing operations, but chiefly are used by the Admiralty for naval purposes. The latest warships have double projectors on one base, as Fig. 295. Large vessels have big single projectors, while destroyers and such-like craft frequently use the pattern shown in Fig. 296.

Shackle Plates, Fig. 297, are simply a base plate with eye and shackle used for heavy standing rigging.

Shackles, Fig. 298, connect the ends of a cable to an anchor, or are used for other similar purposes.

Shrouds, as already shown in the mast drawings, support the main structure of a mast. Their method of attachment to the deck is shown in Fig. 299, which also shows the ratlines, or " ladders," so dear to the heart of the earlier model boat builders.

Sidelights are used to enable other mariners to know at night in which direction a vessel is proceeding. The starboard, or right-

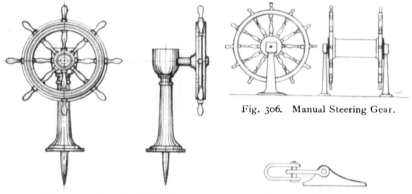

Fig. 306.　Manual Steering Gear.

Fig. 305.　Steering Wheel.　　　　Fig. 307.　Stopper Bolt.

hand side of a boat looking from the stern forwards, is always green, and the left-hand or port side always red. The masthead light is always white. The sidelights are generally carried in a box-like structure, as shown in Fig. 300, which is a good all-round pattern for general use. A larger sketch of a side lamp is given in Chapter XI.

Skylights, as the name implies, are provided on deck to admit light to the spaces beneath. For a working model the simple pattern, Fig. 301, is quite good enough, but a more elaborate fitting is shown in Fig. 302, with seating accommodation, such being suitable for a gentleman's yacht or a liner.

A Sounding Machine, Fig. 303, is used to ascertain the depth of water, and consists of a long line and recording apparatus.

Stanchions are very numerous. Various standard patterns are shown in Fig. 304,

    A being for a single rail, with wood rail top,
    B for single rail only,
    C for two rails, and
    D for three rails.

Their use on a deck side to prevent passengers falling overboard is obvious.

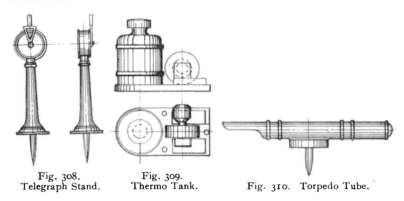

Fig. 308.	Fig. 309.	
Telegraph Stand.	Thermo Tank.	Fig. 310. Torpedo Tube.

Steering Wheels play an important part on shipboard. Fig. 305 shows the general type of metal wheel as used in conjunction with the modern steam steering gear, while the illustration, Fig. 306, is of the manually operated type, frequently used as a stand-by.

Stopper Bolts, Fig. 307, are permanently and firmly attached to the hull structure to provide a sure anchorage for the cables, etc.

Fig. 311. Torpedo Net Poles.

Telegraphs, Fig. 308, of various patterns, single or double handle, serve to convey the captain's orders to the engine room staff.

Thermo Tanks are fitted on a ship to assist in either heating or cooling the atmosphere, and no model passenger boat of any pretence to accuracy can be considered complete without them. Fig. 309.

Torpedo Tubes, Fig. 310, are fitted on deck of destroyers, for the

discharge of torpedoes, while to avoid the disastrous results of being
hit by a torpedo,

Torpedo Nets of strong steel wire are used, held out from the
warship side by the torpedo poles, Fig. 311.

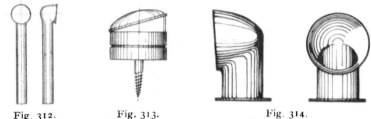

Fig. 312.
Common Ventilator.

Fig. 313.
Cruiser Vent.

Fig. 314.
Destroyer Vent.

Turnbuckles, sometimes called strainers, are useful on large
models for setting up rigging, etc.

Ventilators, as the name suggests, are provided to supply air to
the interior of a boat. A selection of types are illustrated, the
regular mercantile pattern in Fig. 312 ; the naval " light cruiser "

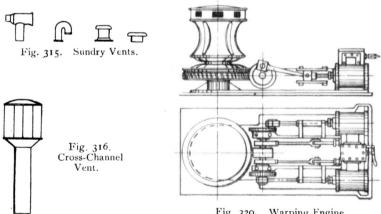

Fig. 315.   Sundry Vents.

Fig. 316.
Cross-Channel
Vent.

Fig. 320.   Warping Engine.

type in Fig. 313 ; destroyer type, Fig. 314 ; sundry different vents
for galleys, seamen's quarters, etc., in Fig. 315. Fig. 316 is a
common form of vent for cross-Channel steamers.

It should be noted that, on modern warships, ventilators of the
ordinary type are conspicuous by their absence, a special system

of forced air supply being used instead, the air intakes being mere gratings or openings in the superstructure.

A *Windlass* is a device for hauling in the anchors, or dealing with any heavy weights. Fig. 317 is the Clarke Chapman windlass, and very extensively used. The " Harfield " windlass, Fig. 318, is

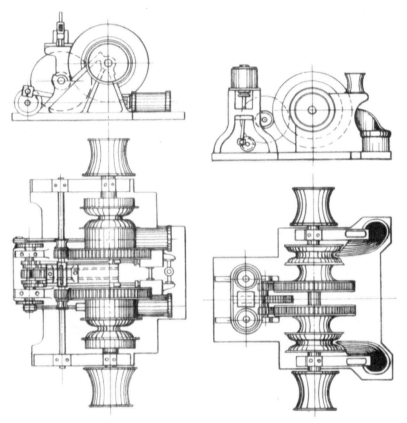

Fig. 317. "Clarke Chapman" Windlass.   Fig. 318.   "Harfield" Windlass.

another well-known type, and has self-contained cable pipes leading to the lockers, also riding bitts cast on to the main structure. A similarly self-contained windlass is the " Napier," Fig. 319, which is a neat fitting in constant demand.

Finally, Fig. 320 shows the Clarke Chapman combined windlass and warping engine, a fitting much favoured for steamers that run a short distance passenger service, and have awkward berths to take.

It has been impossible in the limits of this book to describe

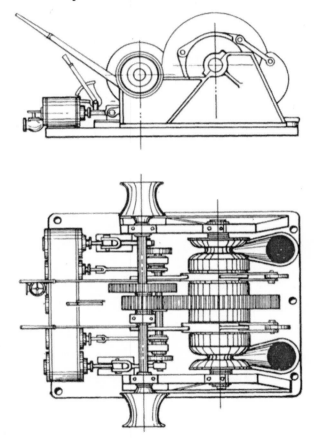

Fig. 319. "Napier" Windlass.

even briefly all the fittings in use, but the foregoing are those most generally needed.

To give some idea of the arrangement and disposition of these fittings on warships and merchant vessels, these further few notes are added.

Taking the typical case of a modern battle cruiser such as the *Lion* or *Princess Royal*, the chief characteristic apart from the big guns is the arrangement of the two big batteries.

The forward battery is shown separately in Fig. 321, from which it will be seen that it is practically a box-shaped erection and may readily be constructed either from thin sheet metal or from hardwood. If the boat is to have electric machinery, hardwood is recommended, but if steam machinery is fitted then the batteries should be made from thin sheet brass or stout tinplates. The dimensions

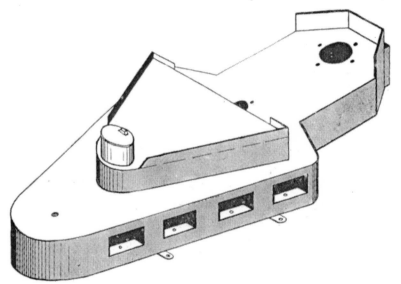

Fig. 321. Forward Battery for Model Battle Cruiser.

can be obtained from the general arrangement drawing of the actual model. The shields around the big central funnel should be specially noticed, and may be cut from the same side plate as the battery. These shields are necessary on the actual vessel to preserve the funnel from the effects of the blast when the big guns are fired, and should not be omitted on the model, as they lend a characteristic finish to the boat. The casemates or gun ports are made by piercing the sides of the battery, and if the latter are made of metal, an angle piece should be attached on the inside of the battery to support the gun, as can be seen in Figs. 321 and 322. If a wooden battery is used, a thin strip of wood should be fitted for a similar purpose. The forward conning tower can be made from a solid block of pine or other suitable wood cut to shape and neatly finished, being screwed into position from beneath. The raised portion of the forward battery should be left open at the back, to admit air to the interior so as to assist in the air supply to the burner, and to help keep the interior of the boat cool. When the forward battery is finished it should be placed in correct position on the deck, and its position

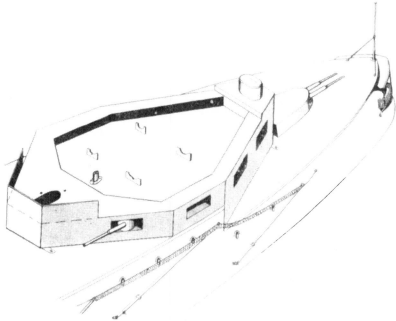

Fig. 322   After Battery and Main Deck of Model Battle Cruiser.

marked thereon, but the whole of the deck from the forward battery and the midship gun to the breakwater at the bows should be cut away to provide easy access to the boiler and blowlamp, and is to be retained in position by fitting a fillet to the underside of the fixed portion of the deck. No special provision is required to keep the removable portion of the deck in place, as its own weight will be found sufficient for the purpose. The after battery, Fig. 322, will be constructed in similar manner to the forward, but it must be noted that from immediately between the funnel and the front of the conning tower a well is formed in which two steam pinnaces and other boats are stowed. This conforms to the present practice in the Navy, as it provides protection for the boats, and reduces the target presented to an enemy's fire. The construction of this portion is a little more intricate than the forward battery, but the experience gained in making the first will simplify apparent difficulties of the latter, but in this case the after battery is to be fitted permanently to the deck and will not be made detachable, as ample access can be obtained to the engines by making only the sunk portion or well detachable.

For this purpose a small fillet or two cross-bars must be fitted to keep it in position. It will be noticed that the after battery fits

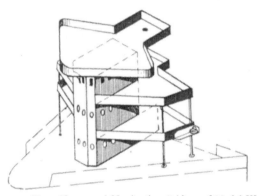

Fig. 323. Chart House and Navigation Bridges of Model Warship.

flush against the ship's side, and when making it should be tried and fitted in position, as, however carefully one works to dimensions, it is almost inevitable that the hull will vary slightly from the original lines. The aft conning tower can be made from a block of solid hardwood in a similar manner to the forward conning tower, and should be screwed in position from underneath, before the batteries are finally fixed in their places, as must be the sixteen 4 in. guns, which should be mounted eight in the fore battery and eight in the after. Fig. 322 also shows the stern walk in place and the torpedo nets, etc.

The chart house, Fig. 323, usually has three bridges, the solid portion, representing the chart houses, etc., being made from a suitable piece of selected timber, the bridges from tinplate, with a vertical strip of metal soldered around the edges to represent the wind screens or dodgers. The windows of the chart house may be shown by painting them with white and picking them out with fine black lines ; or thin metal frames may be cut and fitted in position to show the windows in relief. The circular port holes are then driven into place. The hole for the main-mast should be drilled through all three bridges before it is finally assembled, and it should be noted that the two lower bridges may be secured in position by means of a slot cut in the solid wood block for a depth of about $\frac{1}{8}$ in. and slipping the tinplate into place. From the description it might be assumed that the result will be anything but realistic, but when the boat has been painted it will present a most realistic and handsome appearance, but neat accurate work is essential.

The arrangement of the bow fittings on a modern cargo boat are clearly indicated in Fig. 326, while a typical liner model might be equipped on the same lines. A close study of these sketches will reveal many little points of interest. A typical stern arrangement for a cargo boat or liner, or generally any steam-driven vessel, is sketched in Fig. 327. A usual and satisfactory form of captain's bridge and chart house is shown in Fig. 328, and it is hoped these few rough sketches will assist the reader in correctly arranging the fittings on his model boat.

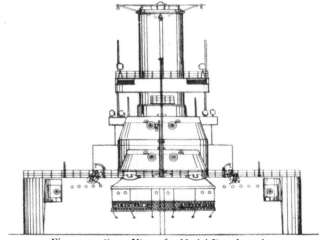

Fig. 324. Stern View of a Model Dreadnought.

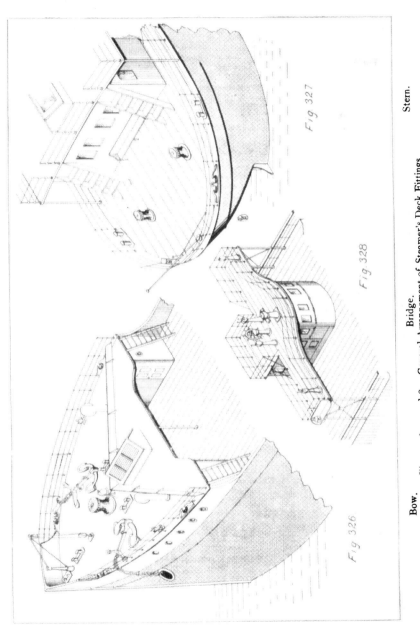

Fig. 327

Fig. 328

Fig. 326

Stern.

Bridge.

Bow.

Figs. 326, 7, and 8.   General Arrangement of Steamer's Deck Fittings.

CHAPTER XIII.

FITTING OUT A MODEL POWER BOAT.

ONE of the most fascinating phases of model boat construction is that generally termed " fitting out," as this expression covers such diverse processes as the painting of the hull, installation of the power plant, and completely finishing the boat, rendering her ready for use on the waters of the lake.

The installation of the machinery into a model power boat of any description is a matter that calls for considerable care and thought. The various parts must all be properly proportioned, suitable to the work they have to do, and almost invariably as light as possible. To achieve these results, one or two important points must be consistently borne in mind. In the first place, the engine or the boiler must not be too large, or one of two serious faults will inevitably follow : either the boiler will be too small to steam the engine properly, or it will be too large for the boat, and the same will have insufficient freeboard, or will be lacking in stability. Similarly in the case of electric models, the motor should be small rather than too large, as best results are obtained with a relatively small motor and a comparatively large accumulator, this combination giving the longest length of run and the most economical working. Bearing these facts in mind, the following tables of standard sizes of engines, boilers and the like have been prepared, and indicate the usual arrangement of boiler and engine under normal conditions ; but of course it must be borne in mind that to secure the very best results the boat, hull, and machinery should have been designed throughout for the special purpose in view, and as detailed in earlier chapters of this book.

It is hoped that the hints in this chapter will also be helpful to those who design and build their own power boat plants throughout.

The actual equipment of a power boat, of course, depends upon the style and arrangement of the machinery and the type of boat being modelled ; but the following hints apply in most cases to power boats, and a few typical examples will illustrate these general principles.   The simplest model with any pretension to realism is a cheap form of torpedo boat destroyer, shown in Fig. 329.

The hull can be purchased ready finished for about 21/- and is of most modern type.   The machinery consists of a simple outfit and is quite easily fitted into the hull by screwing the boiler casing B in position on to a bed formed of a block of wood, fitted to the bottom inside of the hull.   On this chock of wood a thin layer of asbestos is placed, and the boiler screwed into position.   The shaft

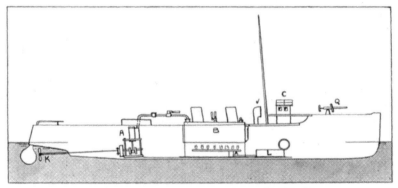

Fig. 329.   Simple Steam Driven Model Torpedo Boat Destroyer.

tunnel, or the hole through the hull in which is fitted the stern tube and propeller shaft K, may be drilled with a long diamond-pointed drill or bored with a red hot wire.   The bush piece provided with the set is then driven into place, and all is made water-tight with some red lead.   The flat piece of metal, also supplied with this set, is to be fitted in-board by bending at right angles and screwing to the bottom of the hull.   The engine A must then be lined up with the shaft, and every care taken to see that the crank shaft is accurately in line with the propeller shaft.   A small block of hardwood will have to be cut to shape and screwed to hull in correct position.   On this block the engine is mounted with ordinary wood screws, passed through the holes already drilled in the engine bedplate.   The deck is then cut to shape and suitable openings or hatchways provided to give access

to the boiler, engine, and lamp L.   The chart house C and bridge
are readily sawn to shape with a fret-saw, built up and fitted in
position.   Guns Q and ventilators V, of simple pattern, merely require screwing into their correct places to complete this part of
the boat.

More powerful models require additional care in their equipment,
especially those with high pressure boilers, and petrol blowlamps,
such a model boat being shown in Fig. 330.

This represents a metre or Class A boat with single flue launch
boiler, petrol blowlamp, and "Simplex" engine.   The engine A
is shown well forward in the bows, and is mounted on a hardwood
block screwed in place. The long propeller shaft requires considerable
care to ensure its absolute parallelism with the centre of the boat,
but otherwise presents no difficulties of construction.   The boiler B

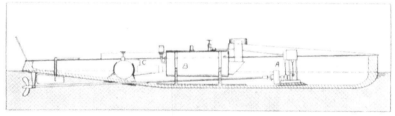

Fig. 330.   Machinery for a Metre Motor Boat.

is mounted on wooden bearers, and secured by metal straps while the
burner C rests upon a shaped wooden block, and is held in proper
position by means of a spring clip, these items being dealt with in
detail further on.

When electricity is adopted as motive power, a much greater
field is available for ingenuity and accuracy in the fitting out and
equipment of the model, which is comparatively simple to fit up, as
the ordinary commercial electric boat motors are usually provided
with holding-down lugs or feet, and all that is necessary is to test
the motor and see that it is in proper working order, and after having
fitted the propeller shaft, cut out a block of hardwood, and screw
the motor into place, taking as much care to see that it is truly in line
with the propeller shaft as in the case of the steam-boat already
described.

The connection between the motor spindle and the propeller shaft is made by a simple shaft coupling or driver illustrated in Fig. 331, and consists of a disc with two driving pins on the same. This is secured to the propeller shaft by means of a set-screw. A similar disc, but with two slots in it, is fitted to the motor spindle, and so arranged that the pins on the disc on the propeller shaft engage with the slots in the other disc.

The accumulator, or accumulators, may be placed in any reasonable position according to the trim of the boat. By simply moving the accumulator forwards or backwards in the hull, when the boat is in the water, the best position is readily obtained. To keep the accumulator in position a light wood box should be made and screwed securely in the hull, and of course a hatch or opening must be provided in the deck to allow of the easy removal of the accumulator. The reversing and control switches may be arranged in any convenient

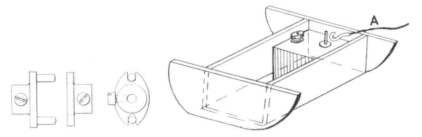

Fig. 331.
Simple Shaft Couplings.

Fig. 332.
Method of Securing Accumulators in Hull.

position on the boat, and connected as shown in the wiring diagrams in Chapter XI.

It is advisable, whenever possible, to arrange the accumulators under a funnel or ventilator to allow the fumes to escape freely into the atmosphere, otherwise the motor and metal parts in the hull will rapidly oxidize and corrode ; and to assist in preserving the metal parts it is always advisable to keep the terminals and other exposed metal parts slightly coated with vaseline.

Having installed the machinery, and seen that the same is satisfactory, attention may be given to the deck fittings, and these of course depend upon the scale of the model, the prototype copied, and one's individual tastes as regards detail.

As it is practically impossible to detail the actual modes of fitting up every type of boat, the following hints have been prepared, and deal in roughly alphabetic order with the leading features, so that the enthusiast may select those that most nearly meet his own requirements, and probably improve upon the methods suggested here. They have all been actually tried in practice by the author, and for their particular purposes are quite good and reliable.

Accumulators must be easily accessible and readily removable; moreover, they must not be easily displaced while the boat is in motion, consequently the simplest plan is to make a cage or box of wood in which the accumulator can rest. This should be well painted, and to save time in connecting the terminals the various wires should be fitted with slotted terminals, clipped on to the insulated wires. These little fittings can be bought at most motor cycle supply houses, such as Gamage's. Another convenience is to

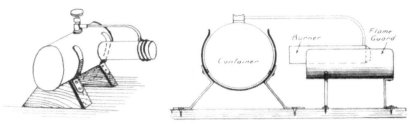

Fig. 333.
Method of Securing Blowlamp.

Fig. 334.
Showing Flame Guard Around Burner.

fix the wires in place with an insulated wire staple, as supplied by any electrician. These prevent the wires dangling about, and getting foul of the motor, etc. An accumulator box, with accumulator, wire lead, and terminal clip is shown in Fig. 332.

Blowlamps.—The essential requirement when placing a blowlamp is accessibility. The best of blowlamps require frequent attention, consequently they should be placed under a hatchway or some readily removable part of the deck. The next requirement is security, and should the blowlamp shift out of place while the model is in mid-ocean, it will in all probability burn the entire boat, or at least do considerable damage. The third requirement, and by no means least, is proper ventilation. A blowlamp induces to a limited extent its own draught of air, but this must be ample, firstly to insure sufficient oxygen for proper and complete combustion of the fuel, and secondly to reduce the local temperature in the boat. In the case of a racer it is customary to leave the blowlamp exposed in an open cockpit, but when the lamp is under the deck, ventilators or a funnel of ample size must be arranged near the burner to insure adequate

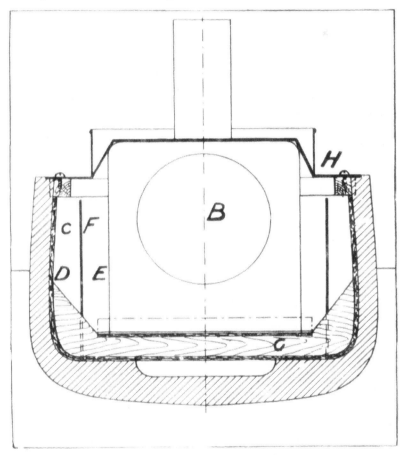

Fig. 335. Fitting Boiler to Hull.

air supplies, and, in addition, any windows or ports that can be cut out should be left open to admit air. A cargo hatch or fiddley grating serves a similar purpose.

To secure the lamp in position an excellent plan is to cut two bearers or a block of wood to shape, and fit it to the hull, on this block attaching a spring steel clip to grip the lamp container, and hold it in place, as in Fig. 333. It is merely necessary in this case to pull the lamp out when it has to be removed. An alternative is to use a light steel or brass cage, as in Fig. 334. To preserve the hull as much as possible, a flame guard, Fig. 334, of thin sheet metal should be fitted around and under the burner tube, and arranged so that when the flame of the lamp is directed downwards while inserting it in the boat, it only plays upon the metal shield. An *air gap* being the best non-conductor of heat, such an arrangement is at once cheap, light, simple and effective. Of course, a thin sheet of asbestos is required on the actual inner surface of the hull all along from the boiler to the blowlamp. Although on purely racing boats this is generally omitted on the

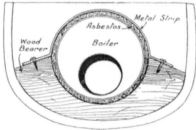

Fig. 336.   Wooden Bearer and Metal Clip for Launch Boiler.

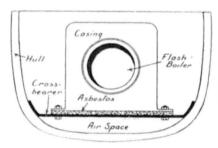

Fig. 337.   Mounting a Flash Boiler.

score of weight, the damage then done to the hull, not being counted as serious, as by so reducing every ounce of weight it becomes possible to win a much-coveted first prize !

Boilers.—These are of two general types :—

      A. With exposed furnaces, such as water tube, Yarrow and flash boilers.

      B. Those with internal or water insulated furnaces, such as single flue, launch, or Scotch boilers.

Those known as exposed furnaces require far more care in their installation, to prevent any risk of burning the hull and blistering the paint on the exterior of the hull, as the only protection from the heat of the lamp is that afforded by the casing and lagging. Fig. 335

shows a water tube boiler of simple type, although the same precautions should be taken with a " Yarrow " or a " Scott " boiler.

The boiler is indicated at B, hull sides at D. A sheeting of asbestos
is nailed to the hull at D, then comes an air gap C, next a thin metal
sheet F, then another air gap E, and finally the boiler casing, which is
lined (or should be) on the interior with a thin asbestos sheet.   To
keep the boiler in place, two hardwood bearers G are fitted as shown,
or two metal brackets could be used.   The wood is preferable, as it
is not such a good conductor of heat as metal.   The boiler is held
in place on top by the metal cross-bar H, insulated from the heat of
the boiler by the thick asbestos pad and the air gap.

In the case of boilers with internal firing, where the burner flame
is entirely surrounded by water, it is obvious that the necessity for
such elaborate precautions is not present ; in fact, a single layer

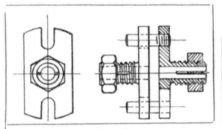
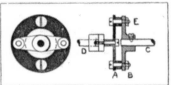

Fig. 338.   Fig. 339.
"Corinthian" Shaft Couplings.   Leather and Metal Shaft Coupling.

of $\frac{1}{8}$ in. asbestos around the outside of the boiler shell is generally
sufficient if an air gap of $\frac{3}{8}$ to $\frac{1}{2}$ in. is permissible ; the boiler is fitted
into bearers of hardwood, and secured with a metal clip as shown in
Fig. 336.

It is not advisable to fix the boiler too solidly, as should the boat
have a violent collision with bank, the impact will unship the whole
plant if only lightly secured in place, whereas if it was too firmly
fixed, the hull itself would in all probability suffer very much more,
as the weight of the boiler and engine would most likely tear out the
wooden foundation blocks bodily from the hull, when considerable
damage would be done to it.

Flash boilers come under the first heading already dealt with,
but as they are generally used on racers, the boiler casing consists
of a simple metal tube or shield, with an under tray or flame guard,

T

and at most a single layer of asbestos on the hull, such as is shown in Fig. 337.

Coupling, between engines and propeller shafts, should always be flexible, as no matter how carefully the various shafts are lined up, they will speedily get out of line, and loss of power would result if flexible couplings were not used. There are many good patterns on the market, especially the "Corinthian" type, Fig. 338. With these no set-screws of any kind are used, but the boss is threaded on a taper and slotted, the hexagonal nut when screwed up closing up the slots in the boss, and so causing it to grip with great tenacity, much in the manner of a drill chuck, a great advantage of this method being the ease with which the couplings can be fitted.

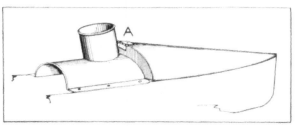

Fig. 340.   Spray Hood, Funnel and Boiler Cover for Racer.

Fig. 339 shows a metal and leather flexible coupling as used on some modern motor cars.   Its advantages are silence and extreme sweetness in operation ; the system being merely to bolt a disc of leather between the two faces of the coupling, while leaving a slight space for rolling motion, due to any inaccuracies of the shaft alignment.   The leather used must be stiff, or the whole coupling will crumple up and fail, or the bolts will tear out entirely.

Decks for racing boats are generally sheet metal plate, the most practical arrangement being the provision of a coned shape spray hood hinged to a boiler casing, provided with a funnel, the boiler casing being secured to the main deck with screws.   The spray hood, which lifts up on the hinge A, Fig. 340, is entirely detachable by removing the hinge pin, while it is held closed with a little spring catch forwards.

In the case of boats with wooden decks and fittings, such as liners, and cargo boats, and similar craft, it is advisable, if the boiler and lamp are adjacent to the deck, to attach a thin strip of asbestos to the underside of deck, to prevent blistering of the varnish.   In all steam-boats suitable openings must be cut to allow the pressure gauge and water gauge, when fitted, to be readily examined, without removing the deck, etc., while the steam from the safety valve should have a free exit to the funnel or a vent, etc.

The regulating lévers to control the steam, blowlamp and the like, should either have extension handles fitted, or be provided with removable turn keys, in the form of a box spanner somewhat as shown in Fig. 341.

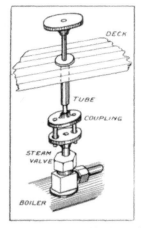

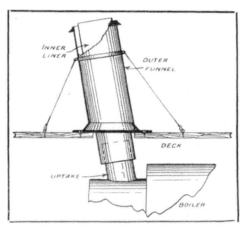

Fig. 341.   Removable Control Key for Steam-boat.

Fig. 344.   Arrangement of Funnel and Liner Tube.

Engines.—The simple types of steam engines merely require screw-ing down with ordinary round-headed wood screws to a hardwood block as already mentioned ; but for higher-class work, proper hold-ing-down bolts somewhat as shown in Fig. 342 may with advantage be used.   These can be made from wood screws cut down and threaded.

For twin screws, or multiple propeller shafts, it is sometimes necessary to gear two or more engines together ; in this case a metal base plate must be provided, an aluminium alloy may be used for lightness.   On this plate the engines, etc., are mounted, while the base plate is screwed to the ordinary wood block.   For very light racing boats a neat arrangement is sketched in Fig. 343, which shows an engine A, mounted on two longitudinal aluminium bearers B,

which have brackets fitted at C to carry the propeller shaft D, the
boiler E being carried on metal brackets M. The blowlamp is in a
cage at F, while at the stern the propeller skeg is fitted as shown at
G, and the rudder post further aft as at H. Thus the entire plant
may be removed bodily from the hull merely by undoing four screws,
and drawing out the propeller shaft. This has a gland at J where it
passes through the hull to keep all water-tight, a rubber packing

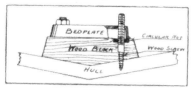

Fig. 342.
Holding-down Bolts for Engine.

Fig. 345.
Spring Catch for Hatch Covers.

under the plate K keeping the joint water-tight. With this
arrangement the hull may be very fragile, as it has but little stress
to take, and merely serves to keep the water out !

Funnels play an important part in the success of a model boat,
quite apart from their scale appearance. For racing boats, these
are but mere rims around the boiler casing, but on other models

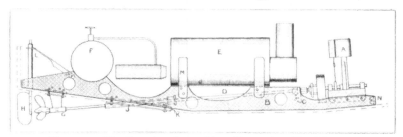

Fig. 343.   The Complete Steam Plant Mounted on Metal Bearers.

a proper internal liner should be provided to protect the funnel
proper from the heat of the burner, somewhat as shown in Fig.
344.

Hatchways must be provided over engines, boiler fittings and blowlamp. Usually a large part of the deck as a whole is arranged to be
removable, or the ordinary commercial form of cargo hatch would
do if fitted over the controlling valves on boiler and engine. There
is no necessity to make any special provision for securing such hatch

covers, although a metal spring catch as Fig. 345 would be advisable on large boats intended for use on the sea.

Lubricating Devices are essential to the well-being of machinery that is in motion ; oil cups with screw lids as Fig. 346 should be provided to lubricate the propeller shaft, engine crank shaft, and other moving parts. Steam chest lubricators are shown in Fig. 347 and Fig. 348, a displacement lubricator in Fig. 349 ; while for boats using forced lubrication a neat brass oil tank,

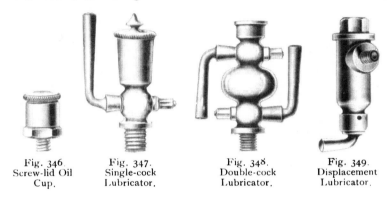

Fig. 346.	Fig. 347.	Fig. 348.	Fig. 349.
Screw-lid Oil	Single-cock	Double-cock	Displacement
Cup.	Lubricator.	Lubricator.	Lubricator.

with lid as in Fig. 350, makes for efficiency and cleanliness, as, of course, any grit or dirt in the oil would prove disastrous.

Motors.—Clockwork motors, as already stated, merely require bedding down on a block of wood, and screwing firmly in position, the only control device being a stop and start lever, which can be arranged in any suitable way such as is shown in Fig. 351, the pad at the end of lever engaging with motor coupling and stopping it when desired.

Electric Motors also present no great difficulties with their installation, a block of wood and two screws being in most cases amply sufficient. When twin screws must be driven from one motor spindle, a gear box is required, such as is shown in Fig. 352. This also is mounted on an extension of the motor block, and is merely screwed in place.

Petrol Motors, by reason of their natural vibration, require very strongly fixing, but the provision of three or four metal brackets screwed to the hull and bolted to the crank-case is usually sufficient, this method being clearly shown in Fig. 353.

Petrol Tanks are best carried in a light metal cage or support, while the coil and accumulator require a neat wood box or metal strap to secure them in position.

Propeller Shafts and *Stern Tubes.*—There are multitudinous ways of drilling the hull, and fixing the propeller shafts, but with a

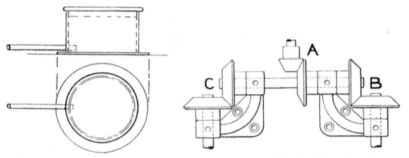

Fig. 350.   Oil Tank for Use with
Forced Lubrication.

Fig. 352.   " Lowko " Gear Box for
Driving Twin Screws.

single screw no great difficulty is found.   Twin screws, however, require much more careful treatment.   There is no golden rule for successful installation of the propeller shafts in their correct position, but the following hints may be of assistance to the amateur shipwright.

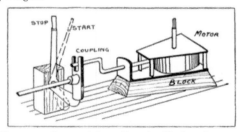

Fig. 351.   Stop and Start Gear for Clockwork Motor.

From the general arrangement drawing of the boat, mark off, on the bottom outside of the hull, that position at which the centre line of the shaft enters the hull, and at the stern mark off the position of the stern bracket.   Next with a long drill, about $\frac{1}{8}$ in. dia., drill holes as nearly as possible in the correct line and angle for the shafts, taking care to see that the point of the drill starts exactly at the right spot on the hull.   This can be done by

first drilling a hole vertically into the hull for a depth of about
$\frac{1}{8}$ in., which allows the drill to be started subsequently at the
correct angle. The two stern brackets may then be fitted in
correct position, and a steel rod, the same size as the propeller shaft
tried in place. The end of this should be heated to a bright red heat,
and carefully worked through the hole, thus burning the hole for

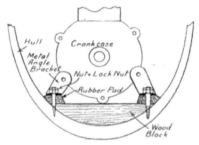

Fig. 353. Holding-down Plates for Petrol Motor.

the shafts ; or if this system is objected to, a long drill with a
diamond point may be used instead, although the red hot wire has
the advantage that one can " feel " that the shaft hole is being
bored correctly and at the proper angle.

As soon as both holes have been made, two perfectly straight steel
shafts should be passed through them, and should also take up a

Fig. 354. The Arrangement of a Stern Tube and Shaft Gear.

bearing in the stern brackets. The boat is next to be placed right
way up and a small cardboard or wooden template, on which are
marked correctly the centres of the engine shafts, prepared and fixed
in place to indicate accurately the positions of the two engine shafts,
unless the engines themselves are already in position.

No doubt some adjustment will be required before these shafts
can be arranged to lie in correct position, and consequently the
holes through the hull will need enlarging until the straight shafts

coincide in the interior of the hull with the engine shaft centres, while the shaft itself is able to turn freely in the stern brackets. The stern tubes are then inserted and fitted to the hull, while a palm piece of flat brass plate must be cut to shape and screwed to the exterior of the hull and soldered securely to the stern tube. To make a water-tight joint, thin wooden wedges should be inserted around the

Fig. 355. The Component Parts of a T.B.D. Stern Tube and Shaft.

stern on the inner side of the hull, and the whole packed with red lead.

The stern brackets are usually built up from strip brass about $\frac{1}{4}$ in. wide and $\frac{1}{16}$ in. thick, on a metre boat, with the bosses brazed or silver soldered on. Soft solder must not be used, or the joints will in all probability give way when power is turned on. The whole arrangement is shown diagrammatically in Fig. 354.

Fig. 356. Stern Tube and Shaft for Boats with Deadwood Sterns.

Stern Tubes are more or less standardised, the essential features being an inboard support or bearing for the shaft, with a stuffing gland or water-tight joint of some kind. A palm piece or support for the tube where it emerges from the hull, and an outboard support or bracket—except in the case of boats with deadwood sterns, like a steam yacht. The arrangement is shown in detail in Fig. 355, the propeller shaft S, stern tube T, gland, palm piece P, and after bracket B being clearly visible. The usual lengths and sizes are given in Table No. 16, although, of course, they must be cut to exactly suit each individual boat.

TABLE No. 16.

STANDARD DIMENSIONS OF PROPELLER SHAFTS AND STERN TUBES.

Length of Shaft.	Dia. of Shaft.	Suitable for
8 in.	$\frac{3}{32}$ in.	30 in. Boats
12 in.	$\frac{1}{8}$ in.	Metre ,,
18 in.	$\frac{3}{16}$ in.	4 ft. ,,
24 in.	$\frac{1}{4}$ in.	$1\frac{1}{2}$ Metre Boats

The above shafts are specially suitable for model racing motor boats, cruisers and such-like craft.

Stern tubes for boats with deadwood sterns are much more simple, the palm piece and after bracket being entirely omitted as shown in Fig. 356. The stuffing glands are removable, and it is simply necessary to drill the hull and cement the tube in place, afterwards soldering the gland in position, the standard sizes being given in Table No. 17.

TABLE No. 17.

STANDARD DIMENSIONS OF PROPELLER SHAFTS AND STERN TUBES FOR MERCHANT MARINE.

Approx. Length of Shaft.	Dia. of Shaft.	Dia. of Tube.
in.	in.	in.
12	$\frac{3}{32}$	$\frac{3}{16}$
15	$\frac{1}{8}$	$\frac{1}{4}$
18	$\frac{3}{16}$	$\frac{5}{16}$

Steering Gear presents many problems from the designer's and user's points of view, but the mere provision of a steering device of sorts on a model is very simple. Several types are illustrated here, and all have their merits. For high speed boats, many practical racing men simply solder a metal blade direct on to the under surface of the hull, and bend it until the best position is found, but a better method is the provision of a fine adjustment device on the tiller as shown in Fig. 357.

This gear is absolutely the lightest and most effective yet produced, but is only suitable for flat-transom boats such as sharpies and hydroplanes.   The tiller, adjusting nut, and plate are of aluminium ; shaft brackets of light brass.   The rudder shaft is slotted and drilled and the blade readily fitted by riveting it into place ; such a device weighing under an ounce, and giving excellent results.   It is fixed by screwing the brackets to the transom, and attaching the screw and plate to the deck.

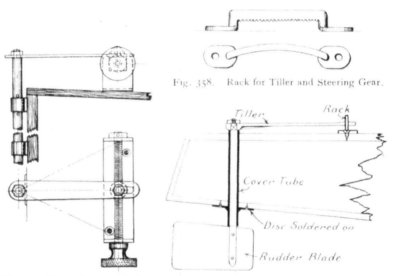

Fig. 358.   Rack for Tiller and Steering Gear.

Fig 357. Fine Adjustment Tiller.   Fig. 359.   Fitting a Rudder to Hull.

A simple rack device is illustrated in Fig. 358, the tiller having a knife edge upon its upper surface which engages with the teeth cut in the underside of the rack.   This form of rudder is usually fitted through the hull as shown in section in Fig. 359.

For spectacular models where it is desired to disguise the tiller, or entirely eliminate it, the quadrant steering device becomes very useful.   This device was brought forward by the writer some years ago and has given splendid results.   The essential parts consist of a capstan, base-plate and socket, pinion and spindle.   Rack with eye substantial enough to bore out to fit any reasonable size of rudder stem. The capstan is mounted on deck, and the rack and pinion is arranged immediately beneath the same, so that by turning the capstan, the pinion is rotated, thus operating the rudder as shown in Fig. 360.

Another method is to use a small steering wheel with bevel gear as shown in Fig. 361, which is self-explanatory.

Whatever system is adopted it is essential that the rudder can be accurately controlled, and securely held in any desired position.

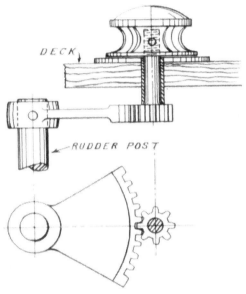

Fig 360. Rack and Pinion Steering Gear.

For speed boats pure and simple, the adjustment should be very fine, and the rudder arranged immediately aft of the propeller, as the column of water thrown aft by the propeller has the greatest effect on the steering power of a boat. Obviously as the power of the propeller jet is enough to drive the boat, surely then it should be sufficient to steer it !—as indeed is the case, except for over-powering external forces, such as heavy wind or, more frequently, sticks and other floating matter in the water. Also no purpose is gained by making the area of the rudder greater than the disc area of the propeller, as this causes the greatest action on the rudder. The action of rudder, propeller and water column is shown diagrammatically in Fig. 362.   An improved rudder and skeg is shown in Fig. 363.

The contention that the best form of rudder for speed boats is narrow and deep, is borne out by the research work of R. T. Hanson and J. C. Hunsaker of the U.S.N. on actual war vessels, and confirmed by practical racing men.

Painting.—The final process in the completion of a power boat is the painting, and frequently is the one phase of ship construction that is most neglected.

When the hull has been properly shaped, sand-papered and finished, it must first be given a coat or two of "priming"; this may

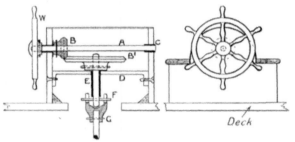

Fig. 361.   Steering Gear for Large Model Boat.

be either a good quality lead colour or a mixture of equal quantities of red and white lead, diluted with turpentine or gold size. This should be applied with a soft brush and well worked into the wood.   Leave this until it is thoroughly dry.   This will take from twenty-four to thirty-six hours according to the time of the year, and it is thus far that the hull should be finished while the machinery and other fittings are added.   When the plant is all correct and in order, the painting proper may continue by rubbing down the surface and giving another light coat of grey or priming paint.   When this is dry, carefully sand-paper it with fine old sand-paper until a very smooth surface is obtained.   If the wood is particularly porous, it may be advisable to give another coat of the priming, or a plain lead colour paint.   This again must be left until thoroughly dry, then sand-paper until it is perfectly smooth.

The wood should now be ready to receive the desired colour.

At this stage either of two methods of finishing the work may be followed. The boat may be enamelled with any well-known enamel, such as Aspinal, Ripolin, Patinol, or Velour, and some excellent results can be obtained from the use of these preparations. Directions for their use are always supplied by the makers, and should be followed exactly, although a few points that should be borne in mind when using the enamel are, to apply a thin even coating of colour, allow it to dry thoroughly hard before lightly sandpapering or rubbing down the surface, and apply a second coat when the first is thoroughly hard, afterwards rub it down with pumice powder until a smooth polished surface is obtained. If a glossy finish is desired, a third coat of enamel should be brushed softly over the surface, taking great care to use a clean brush, to apply

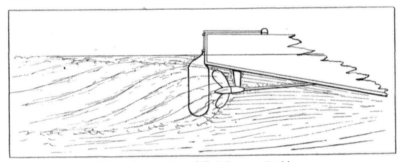

Fig. 362.   Action of Tail Race on Rudder.

the enamel evenly, and above all to work the brush in one direction and not backwards and forwards, as doing so causes the varnish in the enamel to lose much of its brilliance and the excellence of the result would be spoiled. The alternative process is one that gives the very best results, and is obtained by the use of paint known as " coach colour ground in oil." This can be obtained from most high-class oil and colour men, but must not be confused with the ordinary painters' colour ground in oil. Coach colour is much finer, and merely requires thinning down with turpentine to the required consistency, and is applied with a fairly thin brush evenly and lightly to the surface of the boat.

This colour usually dries in two or three hours, but it is not hard enough for polishing for at least twelve to twenty hours, according to the state of the weather. In wet, cold weather, paint takes two

or three times as long to dry hard as in warm weather.   Five or six
coats of this colour should be applied to work up a good body, and
between each coat the work should be very lightly sand-papered,
and in the later stages should be rubbed down with pumice powder.
Of course, while this rubbing down and polishing process is going
on, the boat and brush must be kept well away from dust, and the
work should be wiped down with a damp linen cloth ; afterwards dry
thoroughly with a clean dry cloth before applying the next coat of
colour.   The model should be hung up to dry, deck upwards, in a
room as free from dust as possible.   It is usually feasible to hang any
model power boat in this way, and is worth the trouble, as more dust
settles on to a boat from above than rises up and settles upon the
surface.

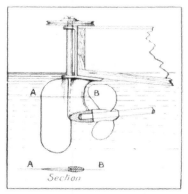

Fig. 363.   Improved Skeg. Shaft Bearing,
and Rudder for Racing Boats.

When the boat has been finished so far with the colour, it must be
varnished, and this requires both skill and practice ; but the best
way to acquire this knowledge is to try for oneself.   A good start
may be made by preparing a small quantity of varnish ; a brand that
gives good results for model work is " Valspar," and this possesses
the virtue of not turning white in the water, as do some of the ordinary commercial varnishes.   Having obtained the varnish, a small
quantity should be poured into a clean tin or lid, and with a small
brush (not too soft) the varnish should be applied, with a true easy
and definite action, evenly over the surface of the model.   The idea
is to put on exactly the right amount of varnish at the start, and not

to be compelled to brush away the surplus from one part to make up the deficit in another. Above all, the brush should be used only in one direction, usually from left to right, and the varnish must not be worked or brushed to and fro, as this causes it to cloud and so lose its pristine appearance. Two or three light coats give better results than one or two heavy coats. In six to eight hours this varnish would appear dry to the touch, but must be left *at least 24 hours* and probably longer before it is rubbed down with pumice powder, as if the varnish is not quite dry and thoroughly hard, the heat generated by the polishing process will " lift " the varnish.

Pumice powder can be purchased ready ground for use and is very much like dirty flour in appearance ; to use it, a pad of soft linen should be made, dipped in water and then dabbed in the pumice powder. This is then rubbed with a short circular motion on the hull, and after three or four turns this circling movement is to be continued and carried forward over the whole surface of the model, dipping the pad in water frequently. Skill can be acquired by practice alone, but the result obtainable is well worth the trouble expended on it. A very beautiful effect is obtained by varnishing the boat with two or three coats of varnish and polishing the same between each coat, very lightly rubbing down the last coat with pumice powder applied by preparing the pad as already described, but covering it over with a single layer of linen. This is made thoroughly wet and used as a rubber. The pumice powder partially penetrates the outer linen covering, thus only very slightly dulling the surface of the varnish, giving it an appearance very much akin to glass. When it is desired to indicate the water-line, or to paint the hull in two or more colours, some means of marking the divisions between the two colours is advisable. This is accomplished in many ways, but the simplest and best is that of employing a scribing block to scratch-mark the water-line, as shown in the illustration, Fig. 364, which indicates graphically the whole process. If a proper scribing block is not available, a substitute may be made with a block of wood, through which has been driven a long metal point.

It will be found that several coats of colour will not obliterate this scratch-line, but with a steady hand and a small brush it is possible to draw a clean straight line defining the two different colours.

To satisfactorily paint a model power boat in a high-class manner requires considerable care and *plenty of time* to insure the result being permanent, but it is well worth devoting sufficient thought and care to the work to insure this result, as it is generally admitted that good paint work on a model sets off the fittings and adds very greatly to the resulting pleasure obtained from the boat.

In the case of liners, and craft with many deck saloons, and such-like structures, it is customary to paint in the windows with Prussian blue, or other colours to choice ; and provided the wood-work has received a coat of priming and a light coat of flat white paint, it is

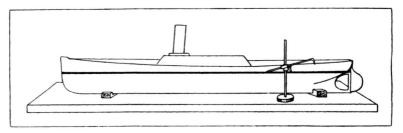

Fig. 364.   Scribing the Water Line as a Guide when Painting the Hull.

quite possible to paint in the windows and so forth with the ordinary water-colour paints as used by artists ; the utmost care is necessary, however, in varnishing over such work, or the paint will be smeared.   A light coat of thin spirit varnish is best unless the boat is to have much work on the water, when, of course, oil colours must be used and varnished over in the ordinary manner.

Exhibition or glass-case models are generally finished and coloured with special spirit colouring media, and finally polished in a manner somewhat allied to French polishing, but it is beyond the scope of this book to detail these processes, and it requires a long apprenticeship to the trade to obtain success.

CHAPTER XIV.

MODEL POWER BOAT RACING.

SINCE the formation of the Model Yacht Racing Association, in 1910, the conduct of model power boat races has been under standardised rules and regulations, and these, coupled with the introduction of the " Restricted Classes," have done more than anything else to place model power boat racing on a high plane, and to give it the dignity of a British sport.

The rules of the Association only provide for certain contingencies and define the course, length of run, method of measurement, and so forth, while the M.Y.R.A. " recommendations " provide for the conduct of race meetings. As this chapter is intended to assist the embryo model power boat racer, these rules can fittingly be included, and are given here to familiarise the reader with the present accepted general practice in the conduct and management of the best power boat regattas.

The M.Y.R.A. Rules for Power Boats provide :—

1. *Management.*—The whole of Parts 1, 3, and 4 of the M.Y.R.A. Racing Rules shall hold good and apply to model power boats, such expressions as "sailing yacht," " sailing," etc., shall be deemed to refer to power boats. These additional or special power boat rules are intended to further guide and direct competitors and officers, and as no rules can be devised to meet every incident of racing and running power models, the Sailing Committee should keep in view the general customs of model power boating, and discourage all attempts to win a race by other means than fair racing and superior speed and skill.

2. *Distinguishing Flag.*—All power boats shall carry, during all races, a clearly defined and prominently placed distinguishing number, which must be so placed and arranged that the same does

not become obliterated or carried away. The numbers may be allotted on the day of the race, or earlier.

3. *Member in Charge.*—Every power boat entered for a race shall be in charge of a member of an Associated Club, either as owner or owner's representative, who shall run that boat only. Verbal advice is permissible, and the owner or representative may have the help of one assistant.

4. *Racing.*—A power boat shall be amenable to the rules from the time the prescribed signal has been given.

5. *Collisions.*—A power boat course shall not be set to involve risk of wilful collision with another boat.

6. *Propulsion.*—A power boat is a boat entirely propelled by

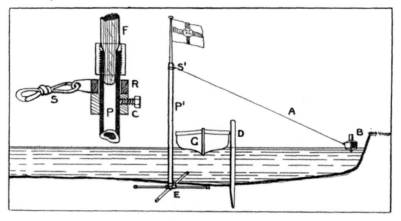

Fig 365. Circular Course Pole.

mechanical means, the whole of the energy being carried or generated in the boat.

7. *Rating Classes.*—The Classes or Rating Rules for National Competition shall be :—

Class A.—The length over-all shall not exceed one metre, beam shall not exceed 20 per cent. of the length, and the displacement in proper working order, with all fuel and water on board, shall not exceed 12 lbs.

Class B.—The length over-all shall not exceed 1½ metres, beam shall not exceed 20 per cent. of the length, and displace.

ment shall not exceed 25 lbs. in proper working order, with all fuel and water on board.

Class C.—The length over-all, excluding rudder, shall not exceed one metre (39⅜ in.). There shall be no limit of beam and displacement.

Class D.—The length over-all, excluding rudder, shall not exceed 1½ metres (59 in.). There shall be no limit of beam or displacement.

Class E.—The length over-all, excluding rudder, shall not exceed 24 ins. There shall be no limit of beam, but the displacement shall not exceed 10 lbs.

Note.—In cases where races are being held on salt water, the boats shall be allowed an extra displacement or weight equal to the difference in density of salt or fresh water.

Note.—Classes A and B are strongly recommended to all club members as the best for all-round work, to insure good sport and to encourage inter-club racing.

Fig. 366.   Steering Competition Target.

8. *Course.*—The course shall, as nearly as possible, be for Classes A, C and E, 50 yards, and Classes B and D, 100 yards. The finishing marks shall be half the length of the course apart.

9. *Straight Course.*—All boats in Classes A and B shall be run over a straight course.

Unrestricted racers, Classes C and D, shall be run on the circular course.

10. *Turning a Boat.*—In a course over 50 yards in length a boat may be turned once only on to her correct course, either by the owner, his representative, or any authorised person, in the event of the boat approaching the bank or other obstacle.

11. *Clearing Course.*—The Sailing Committee shall have full powers to keep the course clear for competitors only, during the race periods.

12.   *Attendance.*—All boats must be reported to the Committee at least half-an-hour before the start of the race.

13.   *Starting.*—All boats shall start at the instruction of the starter, but the propeller must be revolving, under full power, for at least 15 secs. immediately previous to the start.

14.   *Starting Flag.*—All races shall be started by the fall of a white flag, preceded by a predetermined sound signal.

15.   *Order of Starting.*—All boats shall be started in rotation, according to programme numbers.

16.   *Timing.*—All races shall be run against time, and the times shall be taken from the fall of the starting flag to the finish of the course, by means of approved stop watches. All times shall be recorded officially by the scorer, who shall be stationed adjacent to the timekeepers.

At least two timekeepers shall time each course.

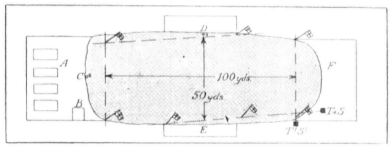

Fig. 367.   Typical Pond Arranged for Racing.

17.   *False Start.*—In the event of any boat failing to start correctly and necessitating a re-start, a notification shall be given to the timekeeper by ringing a bell, or other predetermined signal.

18.   *Re-starting.*—The Sailing Committee shall have power to order a re-start or refuse a re-start (subject to Rule 19) at their discretion.

19.   *Mechanical Failure.*—Any boat which has started on a race and stops from any mechanical cause or failure shall be entitled to one re-start.

20.   *Finishing Line.*—The finishing line shall be clearly defined by means of two poles 5 ft. in height, carrying distinguishing flags at least 2 ft. in the hoist and 3 ft. in the fly.

21. *Obstacles.*—An obstacle is defined as anything which stops a boat other than the sides of the pond, or any authorised person or " catcher." Any boat in a race stopped by an obstacle shall be entitled to a re-start.

22. *Fenders.*—All boats competing under M.Y.R.A. Rules shall carry an efficient fender on the bows.

23. *Lifting Planes.*—No boat shall be fitted with any description of auxiliary lifting plane.

Much of the success of a race meeting depends upon the selection of the officials, and the zeal with which they carry out their duties, but the recommendations that follow indicate the lines usually followed, and although, for example, the timekeepers' job on a wet day is not the most enviable, they at least have the pleasure of " seeing all the fun," and have no need to burn their fingers with refractory blowlamps or scalding hot steam !

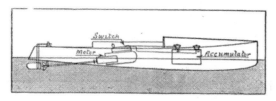

Fig. 368. The Arrangement of Machinery in Electric Boats.

The whole of the ordering of the races, or management of the day's sport, falls upon the officer of the day, and, subject to the rules and regulations, he is in supreme command, as of course, with all well organized events, a controlling official, or head, is really necessary to smooth over any little difficulty that may arise, or to settle some trifling dispute over interpretation of rules, and so forth. The other officers, and their duties are as follows :—

Recommendation for conduct of Model Yacht Races.

A. At the meeting of the General Committee of the Model Yacht Racing Association it was resolved to recommend all model yacht clubs wherever practicable to adopt the following recommendations for the better and more uniform conduct of yacht races ; but the failure of any officer or club to carry out any or all of these recommendations shall not necessarily be deemed cause for dispute or disqualification, as no recommendations can foresee every contingency, but are rather intended to help those less experienced in model power boat racing.

B. *Officer of the Day.*—All the officers at a match shall be under the control of the officer of the day, and he shall arrange the number and disposition of the various other officers, their duties, and so forth ; he shall see that the boats are numbered or marked in the prescribed manner, and that the course, etc., is clearly defined as provided in the rules, and that the scorers, starters, and umpires are provided with scoring and starting cards, etc. He shall be distinguished by a tricolor rosette.

C. *Umpires.*—The duty of the umpires—and two at least should be appointed, one for each side of the pond—are to observe that none of the competitors contravene any of the rules, and that the boats

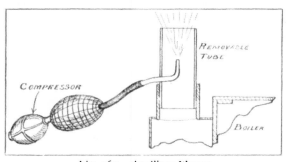

Fig. 369. Auxiliary Blower.

start and finish, and are generally handled, in accordance with the rules. They shall be distinguished by a white rosette.

D. *Scorers.*—There should be whenever practicable two scorers, one at each end of the pond, and they should be provided with a card giving the names and number of the competing boats, and spaces should be left for the score in each heat to be readily marked. They should also be provided with a card giving the " order of starting." In the case of power boat races, they shall record the times as taken by the timekeepers, and in the book or on the card provided. They shall be distinguished by a red rosette.

D. *Starters.*—There should when practicable be two starters, who should be provided with a starting card giving the order of starting, and similar to that used by the scorers. They shall see that the various boats are ready to time and started fairly.

E. *Starting Power Boats.*—The boats should be started as promptly as possible, having due regard to the prevailing wind, etc. The starters shall be distinguished by a blue rosette. Power boats to be started as provided in the rules by the fall of a white flag, the flag being held aloft until the moment of starting, when at the word " go " the flag should be brought smartly to the side to enable the timekeepers to take the times accurately. Immediately prior to this the starter shall blow a whistle, but whenever possible the " flying start " should be adopted by starting the boat from the bank, and taking the times from the time the boat crosses the starting line,

Fig. 370. Return of the Prize Winners.

till she crosses the finishing line. In this case the boat may be started by one appointed person, and the official starter shall indicate the boats passing the starting line by dropping the white flag, and blowing the whistle as already stated.

F. *Timekeepers.*—In power boat racing the taking of the times accurately is of very vital importance, and two or more timekeepers should take all the times with the aid of stop watches of good quality, and if possible synchronized. The times to be taken from the fall of the starter's flag until the boat passes the finishing line in the

prescribed manner.   The scorers shall record the times in a book or
on a card provided.   Timekeepers to be distinguished by a black
and white rosette.

G. *Finishing a Race.*—In power boat races, the scorer and
timekeeper shall together act as judges to decide the order of crossing
the finishing line.   The scorer shall announce the fact by blowing
a whistle when each winning boat passes the line.

The foregoing are the rules generally in force at a well organized

Fig. 371.   A Group of Wimbledon Club Boats and Owners.

power boat regatta, and at present two systems are in vogue for
speed contests :—-

(*a*) The straight course ;

(*b*) The circular or controlled course.

In the first case, the boats are started one at a time, from the bank,
and are timed over a straight course defined by flags.   They are
timed with accurate stop watches, and it is usual to arrange for a
few yards' run before reaching the " starting line," to admit of a
flying start, while of course the finishing line is usually a few yards
from the bank to allow the stoppers or boat catchers to reach the
competing boat without impeding the view for the timekeepers.

The straight course undoubtedly produces a better all-round boat than the circular course, although the latter is easier and better for very fast speeds, as the boat's course is certain.

The circular course is arranged by fixing a vertical pole in the pond, or in a row boat suitably anchored, a convenient height for this pole being 5 ft. from the water level. To the pole a swivel and slip ring are attached and arranged to revolve freely. To these is attached a strong *light* line, well greased, and of known length, usually about 30 ft., although the length of line is usually settled by the geography of the pond, or by a special desire to run the event

Fig. 372. *Scarlet Runner III* at Speed of 16 m.p.h.

over a known length. For example, if the length of the course is to be 100 yds. the circumference of the circle travelled by the boat must be 100 yards, and to obtain this the radius of such a circle would be 15.9 yards, which must be the exact length of the string (from end of boat swivel to centre of pole) to which the boat is attached. A design for a circular course pole and swivel attachment for the boat is shown in Fig. 365. It is important to use a proper plaited and dressed line, well stretched, or serious discrepancies in the length due to the shrinkage of the line will upset all the calculations. In one case under the writer's notice the line shrank over 10 feet during the racing !

Much argument has been heard concerning the effect of centrifugal force on a boat running on the circular course, but its effect, if any, is so slight as to be negligible, as in practice it has been found that the " pull " on the line is very slight. The result of centrifugal force would be considerable if the boat were to rise right out of the water, but any accession of pressure to leeward of the hull, caused by centrifugal force, would tend rather to force the boat inwards or into the circle, and not out of it, as the pressure would probably be ahead of the centre of lateral resistance (C.L.R.) of the hull. Of course, the *line* must be attached to the hull in *front* of the C.L.R., for this very reason, or the boat will run off the course. In addition, there is the weight of the line constantly tending to pull the boat inwards,

Fig. 373.   Running Well.

and this also counteracts somewhat the effect of centrifugal action. The weight of the line also counteracts any tendency to lift the boat by reason of the support given by the line, unless the angle is very considerable.

Steering contests are decided upon somewhat different lines. In this case each boat starts from a fixed point, and steers towards a target of flags fixed conveniently at the opposite bank. The central flag is white and scores a " bull," other flags red and blue, and arranged as shown in Fig. 366, which is the usual arrangement. These contests are very popular, and as it is not uncommon for several boats to score the same number of points, the re-sails so occasioned are always exciting.

The arrangements for the convenience of competitors at a power boat regatta should provide a reserved portion, with tables or boards on which the competing boats may be placed when not in use, and to facilitate adjustments, etc., such an arrangement being shown in the plan Fig. 367, which shows a practical arrangement of a pond for a power boat contest.   Each boat, of course, has a distinguishing number, and these are generally issued by the officer of the day, and are pasted on the deck or other part of the hull. A programme gives a list of the events, with particulars of the prizes

Fig 374.   A Championship Cup Winner.   (Mr Noble's Three Fine Racers.)

(if any), times of starting the different races, and a list of the competing boats with their owners' names, and club racing number. A Primus Stove for heating blowlamps, a reserve supply of petrol, lubricating oil, and a bucket of clean water for boiler filling will be appreciated by the competitors.

There are other contests than those previously mentioned arranged from time to time, by different power boat clubs, but the enthusiast can obtain particulars from the local honorary secretaries, and in the event of any difficulty the editor of the *Model Engineer* would be pleased to assist by giving the names of club secretaries to anyone interested.

It is impossible to give written instructions whereby the amateur
may learn how to run a boat and win prizes at the first attempt.
Most of the successful model yachtsmen have been through a long
apprenticeship by patiently running their boats, and trying this and
that experiment until the best result is obtained.   Moreover, each
boat is somewhat different in detail from all the others, although the
general principles remain the same.   The following hints, although
not given as infallible, or the only method whereby success can be
insured, are at least the result of experience gained by the author's
failures—the best of all schools—and it is hoped will help others
to avoid troubles and disappointment.

Fig. 375.   *Leda IV*, Mr. Vanner's (Victoria M.S.C.) Prize-winning Boat.

The remarks in previous chapters should be considered, and a
general idea of the boat obtained before reaching the pond side ;
in fact, a trial trip in the family bath saves much sorrow at
the pond ! as in the comfort of home it is so much easier to
remedy burnt fingers, and possibly find solace in the bosom of one's
family for the trouble of the would-be record breaker.   Moreover,
an ounce of practice, when it comes to handling a boat, is worth
pounds of theory, and a trial trip at home will enforce upon the mind
the absolute necessity of a supply of petrol for the motor, or blowlamp, oil for the engine, a suitable spanner for undoing nuts, and so

forth, and save many and bitter words when the pond is reached without these useful adjuncts, and it is realised that such things are the very life and soul of the boat.

Model Power Boats and How to Manage Them.—The simplest of all types of power boats are those propelled by clockwork, and for quite young boys, or for use in small ponds or large baths at home, clockwork provides an ideal motive power. They are practically always ready for use, scarcely give trouble of any description, and

Fig. 376. A Trip on the River.

are extremely reliable, the only difficulty being that clockwork cannot be constructed practically to drive a boat much more than three feet long. There are a number of types of clockwork torpedo boat destroyers and battleships on the market, all of which are driven by clockwork, having a large spring set horizontally, the winding spindle in the simpler models usually being found concealed in one of the funnels. No provision is made for stopping the clockwork once it has been wound up, and with these models it is necessary to hold the propeller while winding up the spring, but with the superior model a small lever is provided in the aft part of the boat, to stop and start the mechanism. The spring should be wound up fully, but care taken not to break the ratchet spring by forcing it beyond this point. When not in use always let the spring run down, as it retains its tension better than if left fully wound up. One of the most important items to bear in mind with a clockwork boat is that the power is very limited, and therefore lubrication should be carried out very carefully, using the proper clockwork oil for this purpose. All the gear wheels and spindles should be given a few drops of oil, and a little oil should be put between the turns of the spring itself. A little vaseline rubbed over the flexible drivers will be an advantage, while the propeller shaft should be oiled at both ends, a few drops being placed on the inboard or inside end, and a few drops by the propeller at the outer end. After the boat has been taken from the water it should be wiped dry with a cloth, and a spot or two of oil placed on the end of the propeller shaft to prevent the same from getting rusty.

Electrically Driven Boats.—The advantages of electricity as a motive power for model boats are that it is clean, adaptable, reliable and always available for immediate use, provided the accumulator has been properly charged. Fig. 368 shows the general arrangement of a typical electrically driven boat, and although the type of boat may vary, the system remains the same, and the general instructions given here apply equally to a small model electric motor boat as to a large model electric liner or battleship. First, as regards a supply of current, this is always supplied from an accumulator, or possibly several. The chapter on electric motors should be studied for particulars of the method of charging and storing accumulators. In use, always be sparing of the current, as the motor can only make a certain number of revolutions with the current given out by any particular accumulator. Always see the connecting wires are securely fastened, well insulated, and free from damp, as far as is possible. A few drops of oil on the motor spindle are needed, but the commutator should have only the *slightest trace* of oil, or sparking, with consequent pitting of the brushes, will take place. The brushes should press evenly and firmly but not unnecessarily heavily on the commutator. Keep the switch contacts clean and see they make firm and sure contact, also keep all the accumulator terminals greased to prevent them " furring up " ; if this occurs the green deposit may be cleaned away with strong washing soda and hot water, a remedy to be applied if any accumulator acid is spilt in the boat. Never put the accumulator in the boat without wiping it dry with a cloth, and see the vent plugs are in place, as the fumes from an accumulator cause metal work to oxidize rapidly.

As is the case with all mechanical contrivances, it is imperative to give every care and attention to the machinery of a model power boat if anything like successful results are to be obtained. In the majority of cases it may be taken for granted, when a power boat is

Fig. 377. The Start for the Championship.

bought from a reputable firm, it is in perfect working order, and provided ordinary intelligence is used in the handling of the boat, successful results should follow. Steam boats operate on various systems, which have been explained in other pages of this book, and the simplest, having only a methylated spirit burner and simple boiler, are quite easy to manipulate ; all that is necessary is to see that the container is filled with methylated spirit, taking care not to spill any of this inflammable liquid over any portion of the boat. Two-thirds fill boiler with clean water, oil engine, propeller shaft, and other movable parts, and the model is practically ready for use. The refilling process is merely a repetition of the initial start, and presents no particular difficulties. When the boat is finished with, however, it should be wiped down on the outside with a cloth, and all stains and marks, frequently caused by impurities in the water, removed from the hull, as if these stains are allowed to dry on the outside of the hull they are likely to be very difficult to subsequently remove. The lamp should be emptied of any surplus spirit, and the boiler emptied of water. Engine, boiler, and all parts of the machinery should be wiped down with a slightly greasy cloth, and a few drops of oil placed on the propeller shaft and stern tube gland, which should be turned round a few times by hand to insure the oil lubricating the whole of the stern tube and bearing.

In the case of more elaborate boats, with brazed boilers of the water tube type, the routine when starting may briefly be summarized as follows :—

1. Two-thirds fill the boiler with clean water, this being accomplished by unscrewing the safety valve and filling the boiler by means of a measure or small funnel.

2. Replace the safety valve securely in position after having seen that the spring operates correctly, and is not corroded or jambed.

3. Fill the spirit container, if a methylated spirit lamp. See that the wicks are not too tightly packed in the burner pan. They should be sufficiently close to allow the spirit to rise—in the manner of a sponge.

4. See that the steam valve and other fittings are turned off with the exception of the pressure and water gauges.

5. Light the lamp, and place it in position under the boiler. With some boilers an auxiliary steam raiser, consisting of an air sprayer, or jet in a tube of sufficient size to fit the uptake (or funnel) of the boiler, is advisable ; an india-rubber compressor is then used to generate a current of air in the funnel, and so draw up the fire, much as shown in Fig. 369.

6. In the case of a petrol blowlamp fill the burner in the manner described elsewhere and place it in position under the boiler. Take care in doing so that the flame does not impinge on the side of the hull or boiler fittings.

7. Oil the valves and moving parts of the engine and see that the

same turn freely, also put a drop or two of oil in the propeller shaft, and in the case of twin-screw vessels the gear should be lightly oiled or greased. In a few minutes sufficient steam pressure should be raised to operate the steam blower, which, if fitted, should be turned on as speedily as possible, to obviate the use of the hand-operated compressor, thus causing the burner to work properly; if opened too much, however, steam will be wasted, and a longer time be required to raise the proper working pressure. As soon as sufficient pressure has been raised, steam may be turned on to the engines and they should be helped round by turning the fly-wheel of the engine: this will warm it up and clear away any condensed steam that may have

Fig. 378. Rescue Operations.

accumulated in the valve chest. Drain cocks, if provided, should be opened to allow the escape of any water which may have accumulated. With the smaller models, having plain light boilers, a working pressure of 25 lbs. per square inch will be ample, but with the larger and more powerful brazed copper boilers, 75 lbs. per square inch pressure is usually permissible. As soon as sufficient pressure has been raised the steam should be turned on gently to the engine, and after a few revolutions the same will begin to pick up speed, when the steam may be turned on fully, the blower turned off, and the boat released for her run.

Whenever the boat reaches the shore, look to the lamp, see it is burning correctly, look at the water and pressure gauges, and notice there is sufficient water remaining in the boiler for another run, as of course, if the water is all used up before the lamp burns out, disaster is likely to overtake the boiler, as probably some part of it will be burnt out. As soon as the spirit in the lamp has been exhausted, or the water runs low, the boiler should be refilled with warm water, not cold, as the sudden contraction of the boiler might cause a leak, and the same operations carried out as before. The propeller shaft and driver should be occasionally oiled, and when the run has been completed, after taking the model from the water, it will be found advisable to wipe it dry with a cloth, and also to wipe off any surplus oil or water which may have accumulated in the boat or on the machinery, afterwards putting a spot or two of clean oil on the propeller shafts, both on the inside and outside, and well oiling the engine and valves. This will greatly assist in keeping the whole plant in efficient working order. After considerable use the " Pilot " light on vaporizing spirit lamps will probably require a new wick, but a suitable supply is generally given with the stock boats, and is readily replaced. The safety valve springs and washers should be replaced when they have become worn or leaky, and the springs should be looked at from time to time, to see that the same have not become corroded or stuck up, and also for assurance that same is working properly.

The faster and more powerful boats generally employ a petrol blowlamp and high pressure boiler of the centre flue type, or possibly a Yarrow or one of the other water tube systems. These, of course, call for more attention and skill in their handling, to obtain the best results, but the possibilities in the way of reliability and speed are very much greater than with the " simple " system. Probably the ideal method of driving a steam-boat, for all ordinary purposes, is the adoption of a reliable petrol blowlamp, and the use of the standard single flue launch boiler with a double-action slide-valve cylinder engine. In a previous chapter the system of operation of the petrol blowlamp has been described, and these remarks should be studied in conjunction with the running of the boat, but a few hints may not be out of place.

In the first place, it may be as well to remember that petrol is not explosive until mixed with a certain proportion of air, but it is highly inflammable and its flames are difficult to extinguish ; for this reason care should be taken in the handling of the boat, although a blowlamp when properly handled is undoubtedly the best method of firing a model steam-boat, and, moreover, is perfectly safe in use and operation, so long as common sense and care are always taken. In filling the container, a small funnel with a very fine gauze strainer in the same is advisable; this prevents foreign matter or sediment being carried over with the petrol, and tends to reduce liability for the nipple to clog up when in use ; the container should be about two-thirds filled, and the filler or air valve screwed into place. Lead washers should in all cases be used on the container, or anywhere else in conjunction with the petrol supply system, as they are the only washers that can be depended upon to maintain a tight joint under all conditions, and even these, after continuous use, will have to be renewed from time to time. Previous to this, see that the regulating valve for the petrol between the container and the burner is closed. Then, with the valve still closed, heat the outer end of the burner tube until it is almost red hot, at which stage the air may be pumped into the container and the petrol regulating valve turned on about half a turn, when a very fine spray of petrol vapour should emerge from the nipple with a loud hissing noise. This vapour, mixing with the air in the burner tube, forms a highly combustible mixture, which can be ignited with a match or with the flames of the heating medium. The intensity and heat of the flame is varied by varying the air pressure in the container, and the amount of opening of the regulating valve. The higher the air pressure in the container the more rapidly the spirit emerges from the nipple, consequently the more spirit can be burned (within reasonable limits) in a given time, with a necessarily greater heat energy available.

One of the difficulties, when at the side of the pond, is to obtain a suitable heating source to start the blowlamp. This can be provided, however, by means of an ordinary tin can having one of its sides cut open, into which the burner would be inserted. In the bottom of the can should be placed some asbestos wick, which should be saturated with methylated spirit, and ignited as required. It is possible to use petrol in the same way, but considerable smoke is given from the petrol when burning, and the heat is by no means so intense; moreover, the burner and nipple become badly sooted up, and for this reason methylated spirit should be used wherever possible, as this never soots up the burner, and consequently greatly improves the reliability of the same. To stop the burner in case of emergency, turn off the regulating valve and release the air pressure in the container, as frequently when the burner is hot, and there is a considerable air pressure in the container, a small quantity of gas will continue to pass, even if the regulator is closed. To guard against non-vaporized petrol issuing from the nipple when starting the burner again, see that the regulating valve is kept shut, as should any petrol have been left in the vaporizing pipes, on re-heating the burner the heat will cause the petrol to be expelled in liquid form.

To refill boilers that carry their own water, a tank can be provided in the boat, but the weight of this may be avoided by merely having a suitable tank with a feed pump in same, and a flexible pipe to connect to a check valve in the boiler, this being kept ashore and used as required.

Coal-fired boilers are not often used nowadays, but the foregoing remarks as regards steam raising, oiling of engines, etc., all apply.

Power boats, driven by petrol motors, should always be adjusted before reaching the water-side, when the tank can be filled with fresh petrol, and the motor started. The general difficulty in starting small petrol motors at the pond-side is caused by imperfect vaporization or faulty carburation, due to the varying temperature and humidity of the air at the water-side and in the house. A hot water bottle full of hot water, if it can be obtained, and placed on the carburettor and engine some minutes before the boat is required to run, is frequently a great help in quick starting. With some motors, especially those having a low compression, it is necessary to practically close the air inlets on the carburettor to obtain sufficient suction to cause a spray of petrol vapour to issue from the jet, or over the surface of a wick carburettor ; while with engines having high compression, and consequently a strong suction, a half closed throttle and full open air adjustment gives best results. The tuning up of the carburettor always takes time and patience, but well repays the trouble taken over it. The ignition system is generally reliable, provided a good quality sparking coil is employed, with a fully charged accumulator of sufficient capacity. Note that the sparking plug is sound, all connections clean and good, when satisfactory running should be obtained.

The knack of starting a petrol motor can only be acquired by practice, but remember it is necessary to store up in the fly-wheel sufficient energy to carry the engine over the non-working stroke, consequently the motor crank-shaft should be swung round sharply and decidedly, not exactly "snatched" round, but with a bold vigorous swing.

As before stated, it is quite impossible to give in writing absolutely definite instructions for the management of model power boats, but the author hopes the foregoing hints will be of service to some power boat enthusiasts, and help to make the pastime more successful.

Finally, to win races, remember, it is not always *the boat;* of course, the boat must be a good one, but so must the men, the captain, and the crew! Power boat racing, like all sports, demands of its devotees patience, perseverance, and the attaining of the highest perfection in the hull, machinery, and details of the boat. Always prepare everything well before the race, see the glands and steam joints, etc., are all tight, burner cleaned out, shafts adjusted, and engine oiled, then go in and win, and if you don't succeed at first, well, just try again.

A Model Hydroplane at Full Speed.

CHAPTER XV.

For convenience of readers this Glossary has been compiled, and while it does not claim completeness, it will be found useful in referring to various details mentioned in the text, or frequently found in nautical phraseology.

A

A1.—The highest class obtainable in Lloyds' Register.

Abaft.—A term used to express the relationship of objects on a ship, beginning at the bows. Thus the mast is abaft the bows. Or the expression may be used for an object outside the ship.

Abeam.—In a line at right angles to a boat's length.

Admiral.—The highest rank in the Navy.

Admiral's Flag.—The St. George's Jack, a white square flag with St. George's Cross in red on it.

Admiralty Flag.—A red flag with yellow Anchor (fouled) on it.

Admiralty Mile.—A length of 6,080 feet.

Adrift.—Floating at random.

Afloat.—Borne up and supported by the water.

Aft.—Abbreviation of abaft. The hinder parts of the ship.

Aground.—When the keel of the vessel rests on the ground.

Amidships.—The middle part of a ship.

Ampere.—The electrical term used in reference to quantity of electrical current.

Amperemeter or *Ammeter.*—An instrument to measure the strength of an electrical current in amperes.

Anchor.—An instrument to secure a boat in a certain position when away from a quay or wharf.

Anthracite.—A special kind of hard Welsh coal.

Area of Propeller.—The sum of the actual areas of all the blades.

Astern.—Referring to an object behind a boat—to travel backwards.

Avast.—An order to stop, hold, or cease any operation.

B

Backfire.—An explosion taking place too early on the compression stroke of a petrol motor, causing it to run the reverse way.

Battens.—In general, long thin strips of wood.

Balk.—A hewn log of timber.

Base Line.—In naval architecture, a level line from which all measurements are taken perpendicularly, usually the load water line.

Beam.—A boat's greatest width.

Belay (to).—To make fast a rope.

Below.—General term for the space beneath the decks.

Bend (to).—To fasten one rope to another or to an anchor, etc.

B.H.P.—Brake horse-power.

Big End.—The end of a connecting rod attached to the crank of an engine.

Bilge.—The round in a boat's side, where it commences to take a vertical direction.

Bilge Keel.—Timber or metal strips fitted longitudinally on the bilge of a boat to assist in checking excessive rolling.

Binnacle.—The stand or case carrying the steering compass.

Block.—A pulley.

Bluff (bow).—Blunt as on a barge.

Boat Chocks.—Pieces of shaped wood, upon which a boat rests when stowed on deck.

Bodyplan.—The drawing showing the cross sections of a boat.

Bollards.—Stout timber or metal posts to attach mooring ropes to.

Booby Hatch.—A smaller kind of companion that can be lifted off bodily.

Bow.—The fore part of a boat.

Bower Anchor.—The anchor near the bows and constantly in use.

Brazed.—Two pieces of metal joined together by a fusible alloy of yellow metal.

Brush.—A stationary electrical conductor leading current from or to a rotating conductor.

Bulkhead.—A tranverse division dividing the hull into sections.

Buoyancy.—The quality of floating or being supported by a fluid.

By the Head.—(Of a boat.) Depressed at the bows below her proper water line.

C

Cable.—(a) A rope or chain by which a boat is held at anchor.

(b) A measure of distance at sea, one-tenth of a nautical mile (608 feet), usually taken at 200 yards.

Camber.—The round upon an upper deck.

Canoe.—A small boat, usually sharp-ended, propelled by paddles.

Capping (or *Rail*).—The moulding on top of a bulwark or rail.

Capstan.—A mechanical contrivance for raising the anchor.

Carburation.—The process of mixing atomised fuel (such as petrol) with air.

Carvel.—A boat plank-built, with smooth sides, planks set edge to edge.

Caulking.—Rendering the points between planks tight by forcing in oakum, cotton, or hemp.

Cavitation.—The partial vacuum caused around a propeller blade when revolving at too high a speed.

Centrifugal.—Outwards from a centre.

Chain Pipe.—Pipe on the deck through which the cable passes to the locker.

Chime or Chine.—A longitudinal frame connecting the sides and bottom of a ship, having a sharp angle at their junction.

Circumference of a Circle.—The diameter multiplied by 3.14159.

Cleat.—A shaped piece of wood or metal around which a rope may be fixed.

Clench or Clincher.—A method of boat construction in which the edges of planks overlap.

Clutch.—A mechanical device for connecting or disconnecting two shafts.

Coamings.—The raised edge or frame around the sides of a hatchway or opening through the deck.

Companion.—A structure with fixed sides and sliding roof giving access to the cabins below deck.

Counter.—The projecting part of a boat abaft the sternpost.

Covering Board.—The outer plank of the deck sawn to the shape of the boat sides.

Cubic Measure of Water.—One gallon contains 277.274 cubic inches or 0.16 cubic feet. One cubic foot contains 1,728 cubic inches or 6.233 gallons.

D

D.—Used in naval architecture to denote displacement.

Davits.—Stout iron stanchions or posts with curved arms used for hoisting boats, etc.

Deadwood.—The solid wood attached to the keel either forward or aft.

Derelict.—An abandoned vessel.

Derrick.—A species of crane, used to lift anchors or cargo, etc.

Diameter (*of Circle*).—Circumference multiplied by 0.31831 = the greatest width.

Dinghy.—A small skiff or boat carried on a yacht.

Displacement.—The quantity or weight of water a boat displaces, which weight is always equal to the total weight of the boat when afloat.

Drag.—The increased draft of water aft compared with the draft forward.

Draught or *Draft.*—The perpendicular depth of water a boat displaces.

Drowned (pump).—A pump so placed that the water has free access to the suction valve, and the water surface at a higher level to it.

E

Earth.—An electrical connection for the return circuit made to any part of the machinery frame.

Ebb.—The receding of the tide.

Efficiency.—The mechanical efficiency is the ratio of the power actually available to the theoretical power.

Electro Magnet.—One or more iron pieces temporarily magnetised by an electrical current.

Entrance.—The fore part of a vessel from the cutwater or stem to the part where it swells out to the full beam of the boat.

F

Fairing (a drawing).—A process by which the intersections of curved lines with the other lines of the sheer, deck and body plans of a boat are made to correspond.

Fairlead.—A contrivance through which a rope is led freely and to avoid chafing.

Falls.—The purchases or tackle for hoisting boats on davits, etc.

False Keel.—A piece of metal or timber attached to the exterior of the keel to protect it.

Fathom.—A nautical measure equal to six feet.

Fitted Out.—A boat all ready to use.

Flare.—The outward slope of the side of a boat, from the load water line to the deck-line.

Flat Floored.—Having the outside of hull projecting from keel in an approximately horizontal direction.

Floors.—Strong transverse frames connecting the timbers from side to side with the keel.

Flotsam.—The floating goods from a wreck.

Flush Deck.—A deck having no raised or sunken portions.

Fore Foot.—The foremost part of the keel, where it forms a support for the lower end of the stem.

Fore Mast.—The mast nearest the bows of a boat.

Fore Peak.—The extreme forward part of a boat beneath the deck.

Frames.—The ribs or timbers of a boat.

Freeboard.—The portion of the hull of a boat above the water.

Full (Aft or Forewards). When a vessel is not tapered sufficiently aft or forwards.

G

Gangway.—An opening in the bulwarks or ship side to allow persons to pass to and from the boat.

Garboard.—The plank which is next to and rabbited to the keel.

Gig.—A long boat of four or six oars.

Girth.—The measurement around a vessel from deck edge to deck edge, or other predetermined spots on the hull.

Gland.—A mechanical metal device encircling a rod and used for the purpose of keeping packing material tight in place.

Governor.—A mechanical device whereby the speed of an engine is automatically governed or controlled.

Grating.—An open woodwork construction put in the bottoms of row boats, etc.

Grub Screw.—A headless, pointed screw to fix some portions of a machine.

Gudgeon Pin.—The spindle set across the piston and to which the upper end of the connecting rod is attached.

Gunwale.—A longitudinal wood strip to which the tops of ribs or timbers are attached.

H

Half Breadth Plan.—One that shows water-lines of a boat by halves from the centre line.

Halyards or *Halliards.*—Ropes for hauling up sails, etc.

Hatch Cover.—The removable roof or covering of a hatch.

Hatches or *Hatchways.*—Openings made in the deck of a boat.

Hawse Holes.—Holes cut in the bows of a boat through which the anchor cable passes.

Hawse Pipes.—The pipes or tubes fitted in the hawse holes and leading to the deck.

Head.—The fore part of the boat.

Heave.—To throw. To bring a strain upon a capstan bar, etc.

Heel.—The lower end of a mast, etc. ; the amount of list a boat has.

Helical.—Shaped like a spiral, such as a screw thread.

Helm.—The apparatus for steering a boat ; usually it only refers to the tiller.

Hogged.—(Of a boat.) Higher in the middle longitudinal plane than the two ends. The opposite of sagging.

Hollow Lines.—Horizontal lines of a boat which have inflections.

Hood.—A covering over a hatch or skylight.   The forward curved deck of a motor boat.

Hull.—The body of the ship, distinct from the mast or machinery.

I

Immersed.—In or under the water.

Immersion Wedge.—The portion of a boat put into the water when she heels.

Inboard.—Within the limits of a boat's bulwarks or deck edge.

Inertia.—The resistance of a body to change of motion.

Initial Stability.—The resistance a boat offers at the first movement to being heeled from the upright.

J

Jack.—A Flag.   The Union Jack.

Jettisoned.—Goods thrown overboard.

Jockey Pulley.—An idle pulley used to put tension on a belt or to guide it.

Joggle.—To cut notches in a boat's timbers for the planks to fit into.

Journal.—That part of a shaft which turns in the bearings.

K

Kathode.—(In a cell in which an electrical current passes through an electrolyte.)   The electrode by which the current leaves the cell.

Keel.—The backbone of a ship to which the ribs or timbers, stem, and sternposts are fitted.

Keelson.—An inner keel fitted over the middle of the floors.

King Plank.—The central plank of a deck.

Knee.—L-shaped piece of wood or iron used to strengthen certain parts in a boat.

Knots.—A measure of speed, one nautical mile (1.1515 statute mile) per hour.

L

Labour.—A boat labours when she pitches or rolls heavily, causing her frames to work.

Lateral Resistance.—The resistance a boat gives to a broadside movement.

Laying Off.—The process of making full-size drawings of a vessel from a scale drawing or table of offsets.

L.B.P.—Length between perpendiculars, that is, between the fore side of the stem and the after side of the stern post on deck.

Lee.—The opposite side to that from which the wind blows.

Leeway.—The distance a vessel when under way loses by drifting out of her true course.

Lights.—The navigating lights a boat must exhibit from sundown to sunrise: head light at the mast head, port and starboard on the ship's sides.

Lines.—The general term for the drawing or design of a boat.

List.—To incline.

L.O.A.—Length over all.

L.W.L.—Load water line.

M

Magnetic Field.—A space under magnetic influence.

M.E.P.—Mean effective pressure.

Metre.—A measure of length: 1 metre = 3.280899 feet; 1 square metre = 10.7643 square feet.

Moment.—A weight or force multiplied by the length of the lever upon which it acts.

Mould.—A framework to the shape of a section of a boat.

Moulded.—The depth a timber is made between its curved surfaces.

N

Nautical Mile.—A length of 6,080 feet.

Needle Valve.—A valve of small aperture closed by means of a fine pointed rod.

Negative Pole.—A term for one terminal of a battery, usually painted black.

N.H.P.—Nominal horse-power.

O

O.A.—Over all: extreme length or width, measured over everything.

Offsets.—Measurements taken from centre line of a boat to the intersection of a water line, and transverse section.

P

Painter.—A rope attached to the bows of a boat to make her fast.

Pay.—To run hot pitch, etc., into the seams of a deck after they are caulked.

Pintle.—A vertical pin in the rudder post to carry the rudder.

Port.—The left-hand side of a vessel when looking from the stern towards the bows.

Prime Mover.—Any mechanism for converting energy into motion.

Priming.—In a steam engine, the passage of water with the steam from the boiler to the cylinder.

Poop.—The raised part of a vessel at her extreme after end.

Q

Quarter.—The part of a ship nearest the stern.

Quarter Deck.—The deck abaft the main mast.

R

Rabbit.—A groove cut in the keel, etc., to make a joint with the planking.

Rail.—The timber fitted to the top of the bulwarks or side stanchions.

Rake.—To lean forwards or aft from the vertical.

Reciprocating.—(Of motion.)   Alternately backwards and forwards in a straight line.

Reeve.—To put a rope through a hole.

Ribs.—The frames or timbers of a boat.

Rolling.—Transverse motion of a ship amongst waves.

Rubbing Band.—A projecting metal or wood strip on the boat's side to protect it.

Rudder.—A plate projecting into the water at the stern of a boat, and used to control the direction of motion.

Run.—(Of a boat.)   The narrowing-in of the underwater to the stern.

S

Sagging.—Drooping in the centre.

Scantlings.—The dimensions of all parts used in the construction of a vessel.

Scuppers.—Openings cut in the bulwarks to clear the deck of water.

Seam.—The joint formed by the meeting of two planks.

Sheave.—The wheel in a pulley or block.

Sheer.—The longitudinal vertical curve of a boat's deck.

Sheer Strake.—The uppermost plank of a boat.

Shrouds.—The wire ropes used to set up a mast.

Siding.—The width of a vessel's framing.

Skin Friction.—The resistance of a boat due to the rubbing action on the water.

Starboard.—The right-hand side of a vessel looking from stern towards the bow.

Stern.—The timber at the fore end of a boat into which the planking is fitted.

Stern Post.—A strong timber to which the rudder is hung.

Stern Tube.—The tube passing through a boat's stern post or bottom, and in which the propeller shaft revolves.

Stiff.—Not easily heeled, having great stability.

Stringer.—A longitudinal strip of timber worked into the inner side of the ribs.

Sump.—A small box into which water or other fluids may drip or drain.

Sweeps.—Large oars.

T

Tabernacle.—A strong case or truck to support a lowering mast.

Taffrail.—The top rail where it is continued round the aft side of counter.

Throttle.—A device to regulate the quantity of gas or steam to be supplied to an engine.

Thwart.—A transverse seat in a boat.

Torque.—The force tending to produce rotation.

Transom.—The after transverse board forming the stern of a flat-ended boat.

Tumble Home.—Inward curvature of the sides of a boat above the water line.

U

Unship.—To remove an article from its place.

U Section.—Section of a boat's hull rounded at the keel like a letter U.

V

Veer (to).—To let out or pay out.

Vessel.—A name for any kind of craft or boat.

W

Wake.—A peculiar eddying motion in the water after the passing of a boat.

Water.—One cubic foot of fresh water weighs 62.393 lbs. ; 1 cubic inch of water weighs .03604 lbs. ; 1 lb. weight of fresh water contains 27.7463 cubic inches ; usually reckoned as 27 cubic inches water = 1 lb. weight.

Water Line.—A horizontal plane passing longitudinally through a vessel.

Weather Side.—The windward side of a vessel, or side on to which the wind blows.

Wetted Surface.—The superficial area of that part of the hull which is immersed.

Winch.—A drum with crank handles and pawl used for raising and lowering weights.

Windlass.—A horizontal barrel, revolved by cranks or machinery used for getting in the anchor.

Y

Yacht.—Generally, any vessel which is permanently fitted out and used by her owner for pleasure.

Yawing.—The turning of a boat's head from one direction to another.

A Handsome Model Light Cruiser.

APPENDIX.

TRADE DIRECTORY.

FINISHED MODELS, Castings, Tools, Materials and supplies generally, as mentioned in this book, can be obtained from the following Firms.

The reader is also referred to the pages of the *Model Engineer*, where all kinds of tools, castings, materials, electrical supplies, and finished models are regularly advertised.

A. H. AVERY,
    Fulmen Works,
        Tunbridge Wells, Kent.
                Electric Motors and Supplies.

E. J. BARNARD,
    40, Camden Road,
        Tunbridge Wells, Kent.
                General Supplies.

BASSETT-LOWKE, LTD.,
    112, High Holborn,
        London, W.C.
  Head Office and Works :
Kingswell Street, Northampton.
  Boats, Engines, Boilers, Castings, Finished Parts, and Supplies.

S. BATESON,
    The Bazaar,
        Blackpool.
                General Supplies.

A. W. BOND,
> 245, Euston Road,
>> London, N.W.
>>> *Electric and General Materials, and Fittings.*

CLYDE MODEL DOCKYARD,
> Argyle Arcade,
>> Glasgow.
>>> *Boats, Boilers, Engines, Fittings, and Materials.*

ECONOMIC ELECTRIC CO.,
> London Road,
>> Twickenham, Middlesex.
>>> *Accumulators, Motors, and Electric Supplies.*

GAMAGE'S, LTD.,
> Holborn,
>> London, E.C.
>>> *Boats, Engines, Boilers, Fittings, and Materials.*

T. J. GARDNER,
> Engineers' Supply Stores,
>> Bristol.
>>> *Tools, Materials, and Supplies.*

HAMLEY AND CO.,
> 202, Regent Street, and 89, High Holborn,
>> London, W.C.
>>> *Finished Parts, Fittings, and Supplies.*

W. H. HULL,
> North Western Arcade,
>> Birmingham.
>>> *Finished Engines and Parts.*

W. H. JUBB,
> 29, Norfolk Lane,
>> Sheffield.
>>> *Castings and Supplies.*

Y

KENSINGTON MODEL DOCKYARD,
185, Kensington High Street,
London, W.
Boats, Engines, Finished Parts, and Materials.

W. LITTLE,
Elms Buildings,
Seaside Road,
Eastbourne.
Finished Engines and Parts.

LIVERPOOL CASTINGS AND TOOL SUPPLY CO ,
41, South Castle Street,
Liverpool.
Engines and Boilers, Castings, Screws, and Materials.

RICHFORD AND CO.,
153, Fleet Street,
London, E.C.
Accumulators and Materials.

STEVENS' MODEL DOCKYARD,
22, Aldgate,
London, E.C.
Finished Boats, Boilers, Engines, Castings, and Materials.

STUART TURNER, LTD.,
Shiplake Works,
Henley-on-Thames, Oxon.
Steam and Petrol Engines and Boilers,
Castings, Hulls, Torpedo Boat Fittings.

T. W. THOMPSON AND CO.,
Greenwich,
London, S.E.
Electric Motors, Accumulators, and Supplies.

TURTLE, LTD.,
    North End,
        Croydon.

Finished Parts, Tools, etc.

UNIVERSAL ELECTRIC CO.,
    Great Ducie Street,
        Manchester.

        Accumulators, Motors, and Electric Supplies.

WILES' BAZAAR,
    Deansgate,
        Manchester.

Finished Parts.

WHITNEY,
    129, City Road,
        London, E.C.

        Engines, Boilers, Pumps, Fittings, and Materials.

The " MODEL ENGINEER," published every Thursday, gives reports of regattas, the news of the Model Boat Clubs, and illustrated articles on the latest ideas in model power boat design and construction.

The " MODEL ENGINEER " INSTRUCTION WORKSHOP at 66, Farringdon Street, London, E.C., gives private tuition in model making, and in all branches of light mechanical and electrical work.

INDEX.

Lightning Source UK Ltd
Milton Keynes UK
UKOW01f0621121016

285016UK00001B/7/P